BIM and Big Data for Construction Cost Management

This book is designed to help practitioners and students in a wide range of construction project management professions to understand what building information modelling (BIM) and big data could mean for them and how they should prepare to work successfully on BIM-compliant projects and maintain their competencies in this essential and expanding area.

In this book, the state-of-the-art information technologies that support high-profile BIM implementation are introduced, and case studies show how BIM has integrated core quantity surveying and cost management responsibilities and how big data can enable informed decision-making for cost control and cost planning. The authors' combined professional and academic experience demonstrates, with practical examples, the importance of using BIM and particularly the fusion of BIM and big data, to sharpen competitiveness in global and domestic markets.

This book is a highly valuable guide for people in a wide range of construction project management and quantity surveying roles. In addition, implications for project management, facilities management, contract administration, and dispute resolution are also explored through the case studies, making this book essential reading for built environment and engineering professionals.

Weisheng Lu is Associate Dean and Associate Professor, Department of Real Estate and Construction, Faculty of Architecture, Hong Kong University.

Chi Cheung Lai is a PRC Register Cost Engineer and Director at Northcroft Hong Kong Ltd.

Tung Tse is a PRC Register Cost Engineer at Northcroft Hong Kong Ltd.

BIM and Big Data for Construction Cost Management

Weisheng Lu, Chi Cheung Lai, and Tung Tse

Routledge
Taylor & Francis Group
LONDON AND NEW YORK

First published 2019
by Routledge
2 Park Square, Milton Park, Abingdon, Oxon OX14 4RN

and by Routledge
711 Third Avenue, New York, NY 10017

Routledge is an imprint of the Taylor & Francis Group, an informa business

British Library Cataloguing-in-Publication Data
A catalogue record for this book is available from the British Library

Library of Congress Cataloging-in-Publication Data
Names: Lu, Weisheng, author. | Lai, C. C. (Chi Cheung), author. |
Tse, Anthony, author.
Title: BIM and big data for construction cost management /
Weisheng Lu, C.C. Lai, and Anthony Tse.
Description: First edition. | Abingdon, Oxon: Routledge, 2019. |
Identifiers: LCCN 2018028783 (print) | LCCN 2018038879 (ebook) |
ISBN 9781351172325 (Master Ebook) | ISBN 9781351172318 (Adobe Pdf) |
ISBN 9781351172301 (ePUB) | ISBN 9781351172295 (Mobipocket) |
ISBN 9780815390947 (hardback) | ISBN 9781351172325 (ebook)
Subjects: LCSH: Building—Cost control—Data processing. |
Building—Estimates—Data processing. | Building information
modeling. | Big data.
Classification: LCC TH435 (ebook) |
LCC TH435 .L885 2019 (print) | DDC 624.068/1—dc23
LC record available at https://lccn.loc.gov/2018028783

ISBN: 978-0-8153-9094-7 (hbk)
ISBN: 978-1-351-17232-5 (ebk)

Typeset in Goudy
by codeMantra

MIX
Paper from responsible sources
FSC™ C013985
www.fsc.org

Printed in the United Kingdom
by Henry Ling Limited

Contents

Figures

Tables

First foreword

When Wilson asked me to write the Forewords for his book about BIM, big data, and cost management, I was more than happy to do so. We share the same view about the sluggish development and adoption of BIM over the past decade in this digital era.

BIM itself has moved quite a long way from clash detection to optimisation, and now collaboration. We conveniently describe BIM as 'the Single Source of Truth' in a collaborative working environment. However, there exist some fundamental gaps to bridge before we can apply BIM with ease, such as the SMM used by quantity surveyors. This kind of changes are disruptive, and there is no quick fix solution at this point in time.

BIM and other information technologies are 'disruptive technologies' in the construction industry. They call for a change in the way we work. They demand better collaboration across disciplines and move the resource curve upfront in order to reap the best benefits, especially at the planning and design stage of construction projects.

Integrated procurement for better vertical integration will yield the best benefit for construction and real estate clients. This will demand a further integration of different Information Technology Solutions and Systems. I am one of those keen advocates, who want collaboration not only across professional disciplines but also across software and platforms, such as the successful integration of BIM and GIS (Geographic Information System) by the Housing Authority in 2013. I am a dreamer, and dreams come true with concerted efforts across disciplines. These include integration of BIM, RFID, GPS, smartphones, artificial intelligence, robotics, and the like. Certainly, we should be looking at big data for smart city application apart from green and intelligent buildings at the city scale.

Created by the industry and for the industry, the Construction Industry Council has been the prime mover to coordinate BIM development in Hong Kong. We need to build capacity and capabilities. We encourage Research & Development in BIM and all associated technologies for the betterment of the industry. We aim towards improving productivity, buildability, maintainability, and sustainability. Above all, these process improvements are essential vehicles to help achieve better built quality for a safe and healthy environment on Planet Earth. This is a common goal for our global citizens and practitioners.

I appreciate and applaud the efforts of Wilson and his co-authors. I hope this can motivate more practitioners and researchers to join hands in finding applied solutions for BIM towards our common goal. Let us collaborate to build a better world, and this book can be a catalyst that underscores its significance.

Ar Ada YS Fung, BBS, JP
Chairperson of Committee on BIM,
The Hong Kong Construction Industry Council
21 May 2018

Second foreword

A sea change is taking place in the construction industry worldwide. We have a move towards integrated project delivery enabled by virtual design and construction that is facilitated by the use of BIM. In addition, we have a trend towards collaborative contracting, and the conundrum of this juxtaposition of these concepts is in the integration and explanation of them to the industry at large. This book addresses these issues in a lucid and comprehensive manner, focusing on the cost management process and the quantity surveying profession. The transformational power of BIM and big data is emphasised throughout and the future role of the quantity surveyor speculated upon.

At the same time that China is exercising its influence in developing infrastructure around the world to support its One Belt One Road initiative, the industry is generally moving towards collaborative contracting which is facilitated by BIM. This move is being led by countries such as Australia, US, and the UK. This is a truly global movement and so this book is published at an important time in industry development. Hence, it has a truly worldwide reach even if it is focused on research that has been undertaken in Hong Kong and China.

This volume gives a broad overview of the uses of BIM within the construction industry. Issues such as the use of BIM for cost estimating and management and the impact of BIM on the professions are addressed. The need for change is clearly described in this book. The use of BIM in enhancing productivity and for error checking is described. This is particularly important in reducing rework which is an important driver for improving safety performance. The importance and development of the BIM execution plan is described, and other important issues such as level of development, intellectual property, and the concept of BIM maturity are introduced and explained clearly.

The issue of value for money is always clearly in focus throughout the text, and the roles of big data, data requirements, and data analysis are clearly explained and their uses categorised. A very useful supplement to the book are the case-study vignettes which describe and explain how BIM is incorporated into projects using real-life cases. The way in which BIM enables integrated project delivery is discussed, and the case studies add to the immediacy of this.

This work will be of great interest to students in the field of construction project management and also to the many practitioners for whom BIM has appeared on the scene since they graduated years ago. The fact that the use of BIM is multifaceted and the power of BIM to facilitate integrated project delivery are clearly illustrated in this work.

Steve Rowlinson
Chair Professor of Construction Project Management
The University of Hong Kong
27 May 2018

Preface

About ten years ago, the function of 'automated quantity surveying', together with 'clash detection', has already been placed at the crux of the effort to mainstream BIM in the global construction industry. Numerous studies have been conducted to investigate the technical solutions of applying BIM for cost management (e.g., BIM-enabled quantity take-off, automated generation of bill of quantities, and 5D BIM-enabled cost control). In this text, the terms 'quantity surveying (QS)' and 'construction cost management' are used interchangeably to accommodate the broader readership from both UK and non-UK systems, albeit sometimes there are nuanced differences between the two terms. Professional bodies such as the RICS or the HKIS have sponsored research projects to examine whether BIM is an ally or an adversary for the QS profession. These studies determined that BIM principally benefits QS, but this long-standing profession should reinvent its core competencies. The whole industry seems much relieved. Yet, despite BIM's decade-long maturity, the global construction industry's acceptance and application of BIM in cost management has been sluggish, or at best, not as smooth as it has been expected.

It is against this backdrop that we three decided to write this book, which aims to confront the difficulties of applying BIM for QS, be they technological, attitudinal, cultural, or organisational. Dr. Lu is an Associate Professor at the University of Hong Kong (HKU) and has been involved in BIM research for more than ten years. Mr. Lai is running one of the biggest QS firms in Hong Kong and Mainland China. A fervent member of the BIM community, he is keen to implement BIM in his firm's daily operations to enhance productivity, make QS works less tedious, and improve the image of the profession. Fresh and full of curiosity, Mr. Tse is a graduate of HKU and colleague of Mr. Lai. He leads the firm's strategic BIM development, in addition to his work conducting conventional surveyor tasks. The team thus has a unique blend of flavour by accessing into both academic and practical thoughts. This book is the result of close interactions between us, the author team, and with the industry, government departments, and professional bodies.

This book has ten chapters divided into three parts. Part I consists of three prefatory chapters, which introduce the general background of construction cost management, BIM theories and hard technological aspects, and BIM implementation strategies and other soft aspects. Part II, also three chapters, provides the in-depth analyses of how BIM can best be implemented in construction cost management. Chapter 4 discusses BIM-enabled processes, culminating in a QS-BIM execution plan and its several critical success factors. Chapter 5 provides several real-life cases to further help readers deepen their understanding of BIM for cost management, not all of which are exemplary cases, but their faults, if any, still inform the reader. Chapter 6 examines big data and how it relates to construction cost management. Putting aside the hype factor of this buzz phrase, in reality surveying companies often accumulate excellent big data. This 'buried asset' can be harnessed to better facilitate their QS business. Part III consists of four chapters, which mainly elaborate the challenges and prospects of BIM and big data for QS. Some good practices are recommended, the future of this domain is discussed, and conclusions are drawn.

Although the book is about BIM, big data, and QS, its readership should not be confined to QS-related professionals only. Rather, we encourage all construction stakeholders to participate in the BIM revolution by realising the value of this emerging technology and applying it to enhance our built environment. In this sense, the book is not a conclusion of the past, not a beginning of a new future, but hopefully a small stepping stone leading to more ardent participation from all the stakeholders.

Dr. Weisheng Lu
The University of Hong Kong, Hong Kong
Mr. Chi Cheung
Lai Northcroft Hong Kong Ltd.
Mr. Tung Tse
Northcroft Hong Kong Ltd.
30 May 2018

Acknowledgements

Special thanks go to Ms. Ada Fung and Professor Steve Rowlinson for their enlightening forewords, advice, and continued support. Ada is a past Deputy Director of Housing (Development & Construction) of the Hong Kong Housing Authority. She has been enthusiastically promoting BIM in her huge public housing portfolio and beyond. Praised as the 'Queen of BIM' by researchers and practitioners in Hong Kong, she chairs the Hong Kong Construction Industry Council (CIC) BIM Committee to further promote BIM implementation. Professor Rowlinson of the University of Hong Kong (HKU) sharply spotted the promises of BIM when it was first introduced and has continuously promoted BIM in his teaching, research, and industrial services. He is leading the world's first MSc IPD at Hong Kong University.

We also would like to express our sincere appreciation to Ms. Jing Wang, Dr. Linzi Zheng, Dr. Ke Chen, Ms. Anna Zetkulic, Ms. Nina Niu, Ms. Jinying Xu, Ms. Xi Chen, and Mr. Zhikang Bao for their contributions to draft and edit the book. Conducting research and writing up a monograph like this is exceedingly time-consuming. Without their generous contributions as a team, this book would simply have not been possible. Many thanks to Ed Needle and Catherine Holdsworth from Taylor & Francis for their support and patience. We would like to thank the three anonymous reviewers of the book proposal for their constructive comments.

Our appreciation also goes to Professor Chris Webster, Dean of Faculty of Architecture at HKU, for his mentorship and encouragement, without which we would not have imagined publishing a book like this.

During the course of planning, researching, and writing this book, we have reviewed numerous articles, books, and documents. We would like to apologize if any of them were inadvertently not cited or acknowledged. We also affirm that any mistakes or errors in the book are entirely our responsibility.

Abbreviations

ABW	Activity Based Workplaces
ACD	Association of Construction and Development
AEC	Architecture, Engineering, and Construction
AFUL	Association Francophone des Utilisateurs de Logiciels Libres
AI	Artificial Intelligence
AIA	American Institute of Architects
API	Application Programming Interface
ArchSD	Architectural Services Department
AR	Augmented Reality
BCEGI	Beijing Construction Engineering Group
BIM	Building Information Modelling/Model
QS-BIM	Building Information Modelling for Quantity Surveying
BoQ	Bill of Quantities
BPR	Business Process Reengineering
CAD	Computer Aided Design/Drafting
CBA	Cost/Benefit Analysis
CESSM	Civil Engineering Standard Method of Measurement
CFA	Construction Floor Area
CICRG	Computer Integrated Construction Research Group
CPD	Continuing Professional Development
COBie	Construction Operations Building information exchange
CSI	Construction Specifications Institute
C&D	Construction and Demolition
DBB	Design-Bid-Build
D&B	Design and Build
DL	Deep Learning
DS	Dassault Systèmes
FAQs	Frequently Asked Questions
FM	Facilities Management
GB	Guobiao
GDL	Geometric Description Language
GFA	Gross Floor Area
GIS	Geographic Information Systems

gbXML	Green Building XML Schema
GPS	Global Positioning Systems
GPU	Graphics-Processing Unit
GSA	General Services Administration
HKCIC	Hong Kong Construction Industry Council
HKIS	Hong Kong Institute of Surveyors
HVAC	Heating, Ventilation, and Air Conditioning
ICTs	Information and Communication Technologies
IFC	Industry Foundation Classes
IIT	Illinois Institute of Technology
IP	Intellectual Property
IPD	Integrated Project Delivery
ISO	International Organisation for Standardisation
LCC	Life Cycle Costing
LEED	Leadership in Energy and Environment Design
LLN	Law of Large Numbers
LoD	Level of Development
ML	Machine Learning
MIS	Management Information Systems
MEP	Mechanical, Electrical, and Plumbing
MR	Mixed Reality
NBS	National Building Specification
NEC	New Engineering Contract
NIST	National Institute of Standards and Technology
NRM	New Rules of Measurement
NZS	Standards New Zealand
OA	Office Automation
OCR	Optical Character Recognition
PBO	Project-Based Organisation
PC	Personal Computer
PLM	Product Lifecycle Management
PMBoK	Project Management Body of Knowledge
PMI	Project Management Institute
PPP	Public Private Partnership
QS	Quantity Surveyor/Quantity Surveying
QTO	Quantity Take-offs
RFID	Radio Frequency Identification
R&D	Research and Development
RIBA	Royal Institute of British Architects
RICS	Royal Institution of Chartered Surveyors
SAR	Special Administrative Region
SME	Small and Medium-sized Enterprises
SMM	Standard Method of Measurement
SMM4	Standard Method of Measurement of Building Works 4th edition in Hong Kong

TAS	Takeoff for Architecture and Structure
TME	Take-off for Mechanical and Electrical
TRB	Take-off for Rebar
UCL	University College London
UKCIC	Construction Industry Council of the United Kingdom
UPMC	University of Pittsburgh Medical Center
VDC	Virtual Design and Construction
VFM	Value for Money
VR	Virtual Reality
WebGL	Web Graphics Library
XML	Extensible Markup Language

1 Introduction

From the outset, it needs to point out that the terms 'quantity surveying (QS)' and 'construction cost management' are used interchangeably in this book to cover broader readership from both UK and non-UK systems. The aim of this chapter is to outline the prevailing practices, processes, and problems facing the time-honoured QS profession, as it applies to the global construction industry. It paints a scene through which building information modelling (BIM)'s prospects and challenges for QS can be better understood. A comprehensive description of QS/construction cost management is beyond the scope of this book; readers can refer to numerous other resources for more detailed elaborations. Interested readers already equipped with QS knowledge can skip this chapter or use it as a refresher before venturing to apply BIM and big data for cost management.

1.1 Definitions

The development of the QS profession traces back about 170 years as it relates to the UK architecture, engineering, and construction (AEC) industry (Cartlidge, 2011). In this book, the terms 'the construction industry' and 'the AEC industry' are used interchangeably, albeit the former seems more popular in the UK context, whilst the latter is probably more popular in the US system and more specific in emphasising the three key stages of a project delivery. Professional quantity surveyors (the acronym for which is also QS) with engineering judgement and practical experience are normally taken on board for construction cost management. According to the Royal Institute of Chartered Surveyors (RICS) (1970), the quantity surveyor's role is

> To ensure that the resources of the construction industry are utilised to the best advantage of society by providing the financial management for projects and a cost consultancy service to the client and designer during the whole construction process.

Over the past few decades, the responsibilities of QS have evolved. RICS (2014) states,

> Quantity surveyors are the cost managers of construction. They are initially involved with the capital expenditure phase of a building or facility, which is the feasibility, design and construction phases, but they can also be involved with the extension, refurbishment, maintenance and demolition of a facility. They must understand all aspects of construction over the whole life of a building or facility.

Yet, another definition, offered by surveyor.com and cited by Pittard and Sell (2016), explains,

> Modern QS provide services that cover all aspects of procurement, contractual and project cost management. They can either work as consultants or they can be employed by a contractor or sub-contract.

The importance of efficient cost management can never be overemphasised in construction works, which are often delivered by adopting the organisation form of 'projects' or say 'construction projects'. Cost, time, and quality comprise the top three objectives of construction project management, which are known as the 'Project Management Triangle'. Amongst them, cost is treated as the most important in many if not most cases. As cost overrun invariably besets stakeholders, the management of every construction project centres on improving cost performance.

1.2 Current practice of construction cost management

According to the Project Management Body of Knowledge (PMBoK) (Project Management Institute [PMI], 2013, 5th Edition), project cost management includes the processes involved in planning, estimating, budgeting, financing, funding, managing, and controlling cost so that projects can be completed within budget. Cost management in the construction industry relates to all cost-related activities from project initiation through to successful building occupation and use (Ashworth, 2010). Before construction begins, cost management focuses on cost estimation and cost planning. The objective of the cost estimation is to establish a realistic budget whilst optimising value for money (VFM) for the client. Cost planning aims to develop a pre-agreed cost framework in the most economic manner, whilst cohering with programme requirements, aesthetic considerations, and engineering feasibilities. After construction commences, that focus shifts to cost control and ensuring expenditures are within budget and the pre-agreed cost framework. After construction completes, QS will settle the final payment and final accounts.

Professional QS are usually brought on board from project inception through to settlement of final accounts. QS generally shift their cost-related tasks and responsibilities throughout the project life cycle. A project's life cycle involves several phases from project initiation, to plan and design development, construction, and ultimately handover of the finished product. Table 1.1 summarises the core objectives at each stage proposed by the Royal Institute of British Architects (RIBA).

Table 1.1 Core objectives by project stages

Stage	*Core objectives*
Stage 0 – strategic definition	Identify client's business case, strategic brief, and any other core project requirements
Stage 1 – preparation and brief	Develop project objectives (e.g., project outcomes, quality and sustainability expectations, project budget, and other parameters); prepare initial project brief, feasibility studies, and site condition review
Stage 2 – concept design	Prepare concept design; agree on alterations to inform and issue final project brief; outline specifications and preliminary cost information along with relevant project strategies in accordance with design programme; issue final project brief
Stage 3 – developed design	Prepare developed design; outline specifications, cost information, and project strategies in accordance with design programme
Stage 4 – technical design	Prepare technical design in accordance with design responsibility matrix and project strategies to include all architectural, structural and building services information, specialist subcontractor design and specifications, in accordance with design programme
Stage 5 – construction	Off-site manufacturing and on-site construction in accordance with construction programme and resolution of design queries from site as they arise
Stage 6 – handover and close out	Handover of building and conclusion of building contract
Stage 7 – in use	Undertake in use services in accordance with schedule of services

Source: Adapted from Sinclair, 2013.

Cost management is undertaken in nearly all project phases, whilst its activities can vary widely. Drawing upon the pertinent descriptions drafted by the RIBA and the RICS, the cost management life cycle starts as early as cost estimate in the preparation phase, cost plan in the design stage, preparing tendering documents, cost control throughout the construction stage, and post-tender estimate until the construction project concludes. Figure 1.1 further elaborates QS duties by following the RIBA Plan of Work.

1.2.1 *Preliminary cost estimate*

Cost estimation, the process of forecasting the most likely cost of delivering a project, is perhaps the QS' most important cost-related activity. It forms the basis for other relevant activities, such as tendering, construction contracting, paying back lenders, and revenue generation as early as possible (Southwell, 1970; Skitmore and Marston, 1999; Cartlidge, 2009; Towey, 2012). Cost estimation can be conducted during several pre-contract stages, as summarised in Table 1.2. This section discusses the preliminary cost estimate, whilst the design-stage cost plan and pre-tender estimate will be elaborated in Sections 1.2.2 and 1.2.3, respectively.

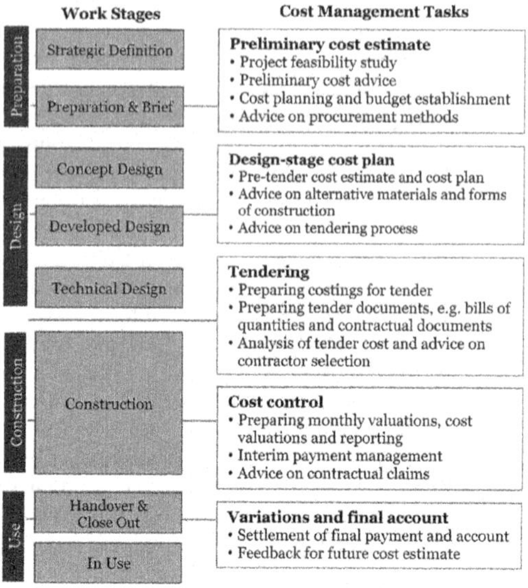

Figure 1.1 QS' tasks and responsibilities summarised in the context of RIBA Plan of Work.

Table 1.2 Estimates types at different project phases

Type of estimate	*Project phase*	*Primary use*
Preliminary cost estimate	Strategic definition (S0), preparation and brief (S1)	Study the feasibility of a project and select the most profitable scheme
Elemental cost plan	Conceptual design (S2), developed design (S3)	Refine and specify the design
Approximate quantities cost plan	Technical design (S4)	Prepare cost plan and tender documents
Pre-tender estimate	Technical design (S4), before tendering	Prepare tender documents

Note: S0–S4 are stages 0–4 defined by the RIBA Plan of Work in Table 1.1.

In the preparation stage when planning for the project strategies and schemes, QS will undertake studies to counsel on the feasibility and profitability of the project. This is usually done based on the preliminary cost estimate, which although quicker to do, usually proves less accurate as little information is available during this stage. Usually, this process considers the land acquisition fee, construction cost, maintenance and operating cost, site servicing, cash flows, market analysis, and revenue and payback period forecast. QS draw on their accumulated experience and recorded cost data to establish a budget. They also source data from past projects of a similar nature or characteristics. If a fixed budget is given, QS may also help to analyse the budget requirements so as to provide guidance on cost planning and available procurement methods. Early cost estimates and budgets are subject to further adjustment once more details become available. Nevertheless, these estimates are of vital importance to the client, helping to inform their decision-making, and even determine whether or not to proceed with the project.

Preliminary cost estimating, also referred to as a top-down, feasibility, or conceptual estimate, still faces limited information being prior to design. Thus, the preliminary estimate relies more on the general project planning parameters, such as project size, type, and location, instead of detailed design or engineering information (Sabol, 2008). QS analyses this information by linking it with the cost information extracted from similar recent projects. Such an estimate method is ease to apply and fast to conduct, but it requires considerable experience and judgement on the part of skilled QS. Some popular methods for preliminary cost estimation include 'floor area method', 'unit method', and 'storey enclosure method' (Akinsiku et al., 2011). 'Floor area method' is one of the most commonly used methods for preliminary estimates. It involves measuring the total floor area of all storeys and multiplying that figure by the historical unit cost appropriated from similar closed out projects. 'Unit method' involves choosing a standard unit of accommodation and multiplying that figure by an approximate cost per unit. It is also known as 'cost according to building function'. 'Storey enclosure method' measures the external walls, floors, and roof areas that effectively enclose the building and multiplies them by appropriate weighting factors to calculate the estimated cost. This method takes many factors into consideration, such as the variations in plan shape, vertical position of the floors, overall height, storey height, and usable floor areas below ground.

1.2.2 Design-stage cost plan

The advice of QS in the design stage can have profound cost implications. As more information surfaces, a more detailed cost estimate could be developed for each functional unit. The weighting of each functional unit against the total cost can also be estimated to justify the economic viability. A comparative cost estimate could be conducted to provide alternatives for high-cost design or materials. If necessary, the change of materials or methods of construction should be pointed out at the design stage to avoid rework in later stages, which bears additional costs. The curves developed by the PMI (2013) illustrate this phenomenon. As Figure 1.2 shows, the later a design change occurs, the more it influences the total cost. It is thus of paramount importance for the client and the designer to work closely with an experienced QS during the design stage in order to control the project's overall cost.

The design-stage cost planning is generally conducted based on the detailed estimating, also known as analytical estimating, bottom-up estimating, fair-cost, or bid estimating, sprouts from detailed design and engineering information. Such estimating is mainly for bid evaluation, contract changes, work scoping, permits and approvals, and often used by the contractor's estimator to perform price extension when the Bill of Quantities (BoQ) become available. When creating an analytical estimate, each individual measured item, such as items listed in the BoQ, is independently analysed by its constituent parts of labour, materials, and plant. Each part is then priced in terms of output, gang sizes, material quantities,

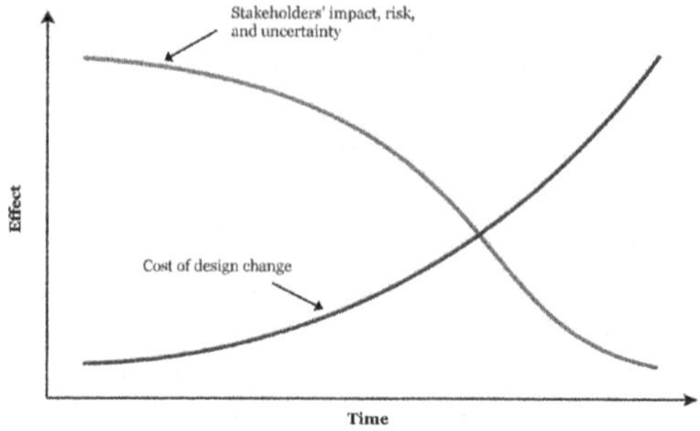

Figure 1.2 Impact of variables based on project time (adapted from Project Management Institute, 2013).

and plant hours. Therefore, the information of the project regarding size, type, and shape holds particular significance. Given the increasing complexity of modern construction and fluctuating material and labour costs, reliable information is difficult to obtain. For example, outputs of labour are subject to variation throughout a single project's construction process. Therefore, in order to cope with subjective analytical estimates, the estimator in practice must use standard outputs, based on individual knowledge, coupled with expected future performance.

The approximate estimating ostensibly generates results that are inferior to analytical estimating. However, when the information is inadequate, the approximate estimating method can help enhance accuracy. Traditionally, designers hand an insufficient amount of information to QS. Therefore, QS principally employ the approximate estimating method. It should be noted, the easier the method of measurement, the more difficult to achieve precise pricing. Perhaps more importantly, associated decisions must be made depending on what information, mainly regarding the quantities and unit prices of all indispensable resources, is at the time at hand to the estimator. Habit, familiarity, and the time available for the estimating process also influence the selected approach.

1.2.3 Tendering

Once the architectural and engineering design drawings are completed, QS will start to conduct pre-tender cost estimation and prepare the tender documents to be sent to contractors/tenderers. When the tenderers return the bidding packages, QS will analyse the tender price and provide advice on the selection of contractors based on their submitted tender documents. According to Cartlidge (2009), the tender documents typically include:

- Two copies of BoQ, one bound and one unbound. The bound copy is for pricing and submission; the unbound copy allows the contractor to split the bills up into trades so that they can be sent to subcontractors for pricing;
- Indicative drawings on which the BoQ were prepared;

- The form of tender – a statement of the tender's bid;
- Instructions (precise time and place); and
- Envelope for the return of the tender.

Specifically, a BoQ includes the quantities of materials, labour work, fees to cover contingency, and any other items that may incur costs during the construction process. A BoQ forms an important part of the contract documents. It not only translates the design documents, such as drawings, plans, and specifications, as reference for potential contractors to base their tender price upon, but also transforms contract documents upon award of tenders, providing an agreement between the client and the contractor.

A BoQ is normally prepared based on the Standard Method of Measurement (SMM). SMM is a uniform industry-wide referencing resource for measuring building works. It provides detailed information, classification tables, and rules for measuring building works to keep the various measuring practices consistent whilst avoiding measuring disputes. Various countries and regions have developed their own sets of SMM for QS' work, including the UK, Canada, the Middle East, Australia, New Zealand, Malaysia, Singapore, Sri Lanka, and Hong Kong to name a few. The SMM, by and large, varies from place to place as each country/region has its own adopted construction methods, techniques, and risk allocation considerations (Hong Kong Institute of Surveyors [HKIS], 2017).

In Hong Kong, for example, the SMM is originated from the UK and lays down a uniform method and criteria for measurement of civil engineering. HKSMM was first published in 1962, but been revised and updated in 1966, 1979, and 2005 to reflect changes in QS practice. Now in its fourth edition, published in 2005, the HKSMM4 is believed to represent "a unanimous industrial agreement on the standard method of measurement", including "Definition Rule (how to define), Measurement Rule (how to measure) and Coverage Rule (what to include and exclude)" (HKIS, 2017). As the increasing adoption of construction and information technology, the HKSMM4 is expected to be updated to reflect these changes and accommodate the existing QS practice in Hong Kong.

1.2.4 Cost control

During the construction stage, the focus of QS shifts from cost estimation to cost control. Cost control is the process of "monitoring the status of the project to update the project costs and managing changes to the cost baseline" (PMI, 2013, p. 215). According to PMI (2013), the main benefit of cost control is to help keep the final construction cost within the client's approved budget by determining the cost variations, evaluating possible alternatives, and taking appropriate actions. In order to maintain effective cost control, the key point is to manage the approved cost baseline and the changes to that base line. An illustration of the baseline is presented in Figure 1.3. One can easily see from the figure that the actual accumulative cost is above the baseline, indicating that the project is performing over

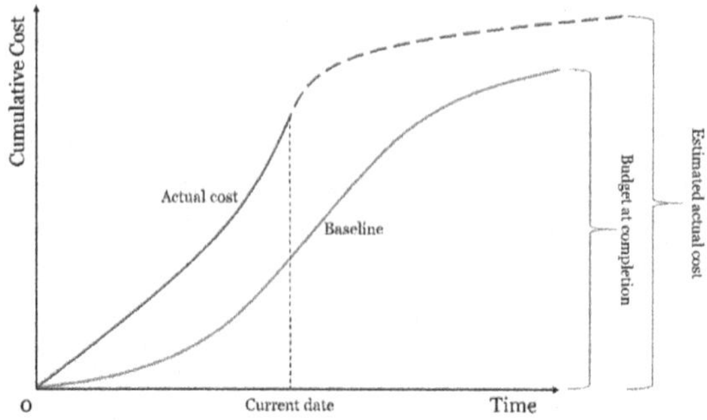

Figure 1.3 Illustration of the baseline and actual cost.

budget, and discretionary interventions should be taken immediately if aiming to complete the project within the approved budget.

Cost control becomes rather important due to increasingly complex construction projects, inadequate planning and preparation at the preliminary stages, and incessant fluctuation in construction costs (HKIS, 2012a). To effectively control expenditure, QS regularly scrutinise their construction project's progress by comparing the actual expenditures with the baseline cost plan and preparing monthly valuations and cost reports for further control actions. Specifically, cost control methods include:

- Preparing an overall cost control plan, which warns of undesirable trends, deviations, slippages, and other project problems;
- Reviewing and approving the work breakdown structure;
- Making and monitoring the cash flow;
- Monitoring and reporting cost during the construction stage; and
- Initiating and approving financial reports to the owner and contractor management.

QS normally call themselves consultants, as the chief architect or engineer would. They are also key coordinators between the client and the client's principal consultant, the lead contractor, or independent project manager, to settle interim payments. The principal consultant will rely on the valuations provided by QS to issue the certificate confirming work completed by the contractor. Only by presenting this certificate to the client can the contractor receive payment for work completed. QS will also manage the evaluation of variations when changes occur unavoidably against the contract documents, as well as to provide advice on contractual claims.

1.2.5 Variations and final accounts

After construction completes, QS will settle the final payment and final accounts in a manner similar to how the interim payment settles. Largely based on the variations between final construction works and the baseline elemental cost plan, this process concludes the QS' obligation to the project. Specifically, the variations could include alterations to the design, quantities, quality, working conditions, or sequence of work. These items could change after the settlement of design to cater technological advancements, statutory changes or enforcement, geological anomalies, unavailability of the chosen materials, and change in market conditions.

Variations should be calculated as early as possible to closely monitor the final contract sum, recognise possible claims, and enable contractors to carry out cost control actions at an earlier stage (HKIS, 2012b). In most cases, contractors will inform QS of any modification in construction works due to architectural or engineering design changes, as well as any other changes identified. QS will also constantly visit the construction site to observe progress and take measurements. Then, QS will further value the variations by quantifying and linking them with the most updated rates. Finally, QS should regularly meet with the contractors to review costs and valuations during the construction and final account periods.

The final account is prepared based on the measured variations, which should be agreed upon by both the contractor and the QS. Notably, according to the contractual requirements, the final account does not necessarily need to be agreed upon by the contractor; however, it is considered a good practice and thus recommended by various QS institutes. To do so, QS and contractors first exchange their assessments prior to the cost review meeting. It is suggested that QS should review the contractors' assessments, mark any inappropriate variation items, and return the commented assessment documents before the cost review meetings. Contractors are also required to go through a similar process. Based on the assessments, QS and contractors exchange comments and try to concur on the variation items to determine the final payment. To facilitate future cost estimation, the cost data could be stored as reference.

1.3 The evolving roles of QS

The traditional role of QS mainly concerns cost management as introduced in Section 1.2. However, another essential role for QS nowadays involves offering legal and contractual advice for contract management. When preparing the tender documents, insurance, warranties, items, and specifications for bond, all must be carefully worded to specify the rights and obligations of the parties involved. Every item in the contractual document shoulders cost implications for both the client and the contractor, thus requiring special attention and expertise from the QS. During the construction process, QS will also provide advice to dispute resolutions if needed. Depending on their professional experience and qualifications, QS can help seek non-court procedures for resolving disputes, such as negotiation, mediation, arbitration, or adjudication, in a satisfactory and cost-effective manner.

Increasingly, the QS profession has seen its roles and responsibilities expanded to cover the measurement of a building's life cycle cost, environmental impacts, and the sustainability of the construction process in conjunction with conventional quantification and cost estimation (Ashworth and Hogg, 2007). As QS are in a position to recommend material selections, they may advocate using more environmentally friendly materials or energy-intensive equipment. Moreover, due to the fragmented nature of construction projects, QS are expected to collaborate extensively with other professionals in the construction industry (Kim and Park, 2016). Contemporary project clients are often critical of traditional QS services and demanding a different and more comprehensive range of services that is proactive, customer orientated, and supported by significantly better management and business skills.

1.4 Problems of existing QS practices

Although QS/construction cost management definitions differ across countries or regions, a systematic QS methodology had pervaded the global construction industry. This QS methodology has helped shape our existing built environment, which is instrumental to human health, economic activity, social behaviour, cultural identity, and civic pride (Pearce, 2003). Nevertheless, existing QS practices are not immune to problems.

1.4.1 *Fragmented process*

The construction industry is notorious for its fragmentation and discontinuity, which have been well recognised by a succession of significant reports into the industry (Latham, 1994; Egan, 1998). The above-introduced practices of construction cost management, following an RIBA Plan, are good for comprehension. However, in real-life projects, projects rarely progress in a strictly sequential manner. Very often, incomplete design or design without sufficient details is passed to QS for estimating and tendering. Even during the construction stage, design changes are common. Although not all design changes prove negative, these changes unavoidably complicate QS work.

In the current way of working, the developer or the client typically signs separate contracts with its designers, consultants, and contractors. As a consequence, the professionals do not always work together efficiently and, in fact, can have competing interests (MacLeamy, 2008). Different professionals, including surveyors, can end up working in silos. There are many calls for the mitigation of fragmentation and discontinuity via technological and organisational innovations. BIM and Integrated Project Delivery have popularised over recent years as a possible option.

1.4.2 *Estimating 'the inestimable'*

Accuracy is the lifeblood of QS services. Given adequate information (e.g., design of sufficient clarity and detail), an experienced QS is able to provide accurate cost estimate. However, in real-life construction projects, QS professionals rarely receive design with sufficient details and specifications. During the conceptual design stage, when designers consider different design options, which normally exist as conceptual ideas only, they need the QS' cost advice. The information availability at this stage does not allow an accurate estimate as many decisions affecting cost (e.g., materials or project timeline) have yet to be made. Even at the developed or technical design stages (see Figure 1.1), the design can be too rough for QS to develop a precise quantity estimate. They may have yet to consider potential design ambiguities or clashes, and as the project progresses, unavoidable design changes can further complicate the issue.

However, that is the 'life' for QS. They seemingly have to estimate the 'inestimable'. There are calls from QS to demand a single truth model based on which stakeholders like clients, designers, and of course surveyors can have a shared understanding of design. For the most part though, QS accept they work with imperfect information. Rather than throwing the problems back onto designers and accuse them of providing poor information, QS provide reasonably precise quantity, cost, and price information to their clients based on their experience, accumulated data, and technical tools. In some cases, particularly when there is quality big data, QS can provide more accurate estimates, which are desired in every cost aspect throughout the project life cycle.

1.4.3 *Tedious work*

One of the major tasks performed by QS involves quantity take-offs (QTO). The conventional way to take-off quantity is to conduct manual measurements based on 2D printed drawings. The measurement subsequently transfers from dimension paper (dim paper) to spreadsheets and BoQ to allow cost estimation. However, the conventional manner of QTOs has long been criticised as laborious, time-consuming, and error-prone (Wong et al., 2014b). Studies by Hannon (2007) and Mitchell (2012) discovered that QS spend 50% to at times 90% of their working hours on quantification. It is not uncommon that QS used all kinds of mark pens to mark the elements on the drawings and put them on the dim papers to form the BoQ. Owing to the size of contemporary projects, this measurement work often involves massive manual works. One needs to pay great attention to detail. Yet, the process is still error-prone. There are calls for the development of various automatic quantity generation software that can rescue QS from these tedious works.

1.4.4 *Time pressure*

Although this has not been documented in literature, it is an enigmatic, widespread observation that every stakeholder (e.g., clients, architects, engineers, surveyors, contractors, and subcontractors) feels acute time pressure in construction business. QS in particular are often under big time pressure. For example, after the designers (i.e., architects and engineers) approve the design, it requires QS to quickly develop the BoQ and other tendering documents for tendering. The time left for the main contractor to respond/bid is also very limited. Normally, the main contractor further pushes its subcontractors in preparing the individual bids. There is a widespread anxiety amongst the QS professionals. Working overtime is becoming a natural part of a QS' life.

There are calls for the reuse of surveying firms' accumulated big data in responding to the ever-heightened time requirement in providing QS services. Companies normally accumulate good data about the quantities of numerous items such as doors, windows, slabs, or walls. A QS can reuse the data in their

QTOs without necessarily starting from the scratch so that the time to undertake cost estimate can be reasonably shortened. In addition, the Law of Large Numbers suggests that the average of the results obtained from a large number of trials should converge to a certain value as more trials are performed (Sen and Singer, 1993). That means, some of the quantities of the items could be more converged and precise if there is big data. By mining the big data accumulated by a company, it is possible to provide quicker and more precise estimate in QS services.

1.4.5 Image problem

Poor image, such as the reputation for underperformance and low-quality working practices, has been a serious problem troubling the construction industry (Escamilla et al., 2016). This problem is even more serious for QS. There is a stereotype that QS are wearing thick eyeglasses, boring, and old hat. QS are often misunderstood as doing routine monotonous work, despite the fact that they also provide a lot of high-profile and creative services to their clients. The status of QS has been somewhat dwarfed by other built environment professions. Often dismissed as secondary to the work accomplished by architects, engineers, and builders, QS gets typecast as the drudgery side of a building project, void of the creativity of the design side, and the physicality of construction.

The image problem seriously endangers the development of this professional group, since the young generation would be reluctant to work as QS in the construction industry. In order to recruit and retain skilled labour force, the image of QS should be improved. One possible and promising strategy for image improvement could be the use of advanced digital technologies, such as BIM, big data, and artificial intelligence (AI), for example, for automatic quantity generation and for rescuing QS from tedious work.

1.5 Summary

Cost is a definitional criterion to the success or failure of any construction project. The importance of efficient cost management cannot be overemphasised in construction works. Over the past century or so, QS/construction cost management has evolved into a long-standing profession with a systematic methodology, a set of rigorous practices, and a clear definition of its core competencies. However, despite the success it achieved, QS has also received its fair share of criticism. Existing QS practices in part add to the fragmentation and discontinuity of the AEC industry that should be mitigated through technological and organisational innovations. QS are often doing difficult works, which, however, are often despised by other professions involved in AEC. They have to provide accurate estimates even when the information is not readily available. They have to take tedious work such as QTO and measurement as a part of their professional life. They often work under serious time pressure. Despite its indispensable, value-added services, QS has an image problem. The long-standing profession is standing at a crossroads, facing greater challenges and increased expectations than ever.

2 BIM theories and technologies

The aim of this chapter is to provide some general descriptions of BIM, including its related theories, key technological developments, information format, and BIM standards in various economies. First, it tries to demystify the concept of BIM, often plugged as a cure-all for the construction industry's woes. An overview of BIM software programs and their potential extensions to other built environment-related activities are subsequently elaborated. It then attempts to articulate the 'information', as the keyword of BIM. The attributes of the 'information' are further discussed in the Level of Development (LoD), which provides common terminology to help users understand BIM information and the degree to which it can be relied upon. The chapter goes on to introduce BIM standards in various economies. Bearing in mind that the book is about BIM, big data, and cost management, the chapter then introduces how BIM libraries can store information, including cost information. It is not the intention of this book to provide a comprehensive elaboration of BIM. However, we hope that readers may find this chapter useful in providing some fresh perspectives towards BIM, which has been somewhat mystified.

2.1 What is BIM?

Growing interest in BIM on the part of business executives, building managers, design professionals, policy-makers, researchers, and the like has extended BIM's presence in the global AEC industry. Various governments around the world see it as a strategic development and seek to mandate its use in public projects. Professional bodies try to understand the emerging domain and embed it as an indispensable part of their professional competencies, whilst companies rush to grasp and adopt BIM for fear of losing their competitive edge. In academia, keynote speakers and international conference participants frequently cite BIM for their subject matter, hailing it as the future platform for AEC. BIM is even advocated as the disruptive development that will bring a paradigm shift to the AEC industry. So, given that BIM is credited with inciting paradigmatic change in the global AEC industry, it is important to appreciate what BIM is and is not.

According to the National Building Information Model Standard (NIBS, 2015),

> Building Information Modelling (BIM) is a digital representation of physical and functional characteristics of a facility. A BIM is a shared knowledge resource for information about a facility forming a reliable basis for decisions during its life-cycle; defined as existing from earliest conception to demolition.

Chuck Eastman, a leading authority on BIM based in the Georgia Institute of Technology, defines BIM from an interoperability perspective as

> A digital representation of the building process used to facilitate the exchange and interoperability of information in digital format.
>
> (Eastman et al., 2011)

The early definition of the term interoperability applies to investigating the issues of information exchanged in the field of information technology and systems engineering services. From a purely technology-based view, interoperability concerns the ability to manage and communicate electronic product and project data amongst collaborating firms (McGraw-Hill Construction, 2007). Nowadays, interoperability refers to the capability of cooperation between systems and organisations such that they can work together (Chen et al., 2008).

Eastman et al. (2011) further suggest,

> BIM is a verb or an adjective phrase to describe tools, processes and technologies that are facilitated by digital, machine-readable documentation about a building, its performance, its planning, its construction and later its operation.

Gu and London (2010) define BIM as

> ...an IT-enabled approach that involves applying and maintaining an integral representation of all building information for different phases of the project life cycle in the form of a data repository.

BIM is thus a polysemous word. It can stand for the digital representation of the physical and functional characteristics of a facility (Building Information Model), or the process of making the model and further harnessing its data (Building Information Modelling). This resonated with Davies and Harty (2011), who suggested that

> BIM has become a common nomenclature to refer to a family of technologies and related practices used to represent and manage the information used for, and created by, the process of designing, constructing and operating buildings.

Besides the aforementioned BIM definitions, a dozen other definitions exist. The proliferation of BIM definitions speaks to the fast growth of the domain. It also signals the potential for confusion to arise when ill-defined terminology is employed to demonstrate specific ideas.

It would constitute a misperception to consider BIM an advanced form of Computer Aided Design/Drafting (CAD). Although BIM is indeed enjoying the kind of momentum CAD experienced in the late 1970s, the two systems differ widely. Traditional 2D CAD drawings are sets of lines, polygons, or surface areas. Whilst 2D drawings can be assigned with dimensions (e.g., length), they are so to speak 'dead' graphics. In contrast, BIM is 'live'. A 'richer repository' of information (Eastman, 1999) than a set of drawings or static CAD files, BIM has the ability to store different types of information.

Although somewhat of an oversimplification, developing a BIM resembles assembling LEGO pieces, the colourful interlocking plastic bricks produced by LEGO Group. When constructing a BIM, a component is selected from a library of components and pieced together with subsequently selected components. When a designer changes a component, its dependent components change accordingly. The BIM model details the properties of each component, such as its material and quantity information. BIM thus has the potential to automate measurement and facilitate the preparation of accurate estimates (Hannon, 2007). Component information is expandable to the extent that it can contain such things as carbon emission figures, thermal properties, and associated construction waste. Therefore, BIM has the potential to calculate carbon footprints, spur energy-efficient design, and manage construction and demolition waste.

2.2 Overview of commercial BIM software

To reiterate, BIM is a digital representation of a facility that is to be built, is being built, or has been built (i.e., the latter two define as-built BIMs). The digital representation is materialised in BIM software packages/solutions. Many people mistakenly equate BIM with such software packages/solutions themselves. A suitable analogy could be typing an electronic document in Microsoft Word. Word functions as the platform from which one sees and produces the document. Word (i.e., the software) is not the same thing as the document (i.e., the BIM). As BIM grows in popularity, more and more software packages calling themselves BIM software enter the market. According to the Illinois Institute of Technology, there are more than 30 BIM programs, and each serves different purposes. The subsequent section briefly describes the common BIM software, but does not indicate any personal preference on the part of the authors of this book.

2.2.1 Revit and Navisworks

Revit is probably the best-known BIM software and current market leader in architectural, structural, and mechanical, electrical, and plumbing (MEP) design. A start-up company originally developed it as an architectural design and information documentation platform for parametric 3D modelling. Acquired by Autodesk in 2002, Revit began to pervade the AEC market. Multiple specialised versions were released over the coming years to enable BIM use for specific disciplines (e.g., Revit Structure in 2005 and Revit MEP in 2006). These various disciplines were then rolled into one product under the blanket *Autodesk Revit* in 2013, an integrated BIM platform that facilitates collaboration amongst different stakeholders.

Revit aims to provide a user-friendly platform for a quick BIM process and analysis based on the BIM. It not only supports the creation of predefined and parametric BIM objects but also provides embedded rules (e.g., constraints on distances, angles, and the number of objects arranged) to ensure the models form realistically. These rules, by and large, shape the hierarchical relationship of various BIM parameters. In this regard, an object in Revit can comprise a class of sub-objects with parametric connections. Revit also provides bidirectional associativity between drawings, models, components, views, annotations, and schedules, making it much easier to edit the model and extract its information. The final products of Revit can be viewed, modified, and stored in a number of file formats such as .dwg, .dwf, and .rva.

Also included in Autodesk's BIM product package, Navisworks is a project review software that facilitates a comprehensive appraisal of aggregated models and data. Navisworks animates for architects, engineers, contractors, and other stakeholders the construction process over time (i.e., 4D simulation) and any detected clashes, underscored by colour. Navisworks can merge different 3D models, source comments and mark-ups, measure distance, and export photorealistic rendering.

2.2.2 Bentley Systems

Bentley Systems incorporate a series of functions to provide a full cover for the business of architecture, engineering, infrastructure, and construction. Expressly launched for BIM design purposes in 2004, Bentley builds up a good standard for the predefined parametric objects, allowing the predefined parametric objects to be extended via the MDL Application Programming Interface (API). The built-in Parametric Cell Studio module support customised parametric objects. Equipping the B-spline surface and strong modelling capabilities reinforces 2D detailing and annotation on a 3D modelling. The solid drawing capabilities of Bentley makes it simple to showcase the actual line weight and text, as well as modify the properties of object classes. The drag-over operation hints, smart cursor, and the user definable menu setups provide a customised interface in Bentley.

The operations conducted in Bentley are built upon file-based systems such that all actions documented in a file can quickly respond to the system. In this regard, Bentley helps the computer shoulder lower loads on memory and thus run more efficiently. Although the versatile features of Bentley Systems can support nearly all aspects of the AEC industry, its strengths particularly suit civil engineering projects.

2.2.3 ArchiCAD

First introduced by the company Graphisoft in the 1980s, ArchiCAD has a long history of supporting architectural design. After 2007, a German CAD organisation acquired Graphisoft, as result of which ArchiCAD applications lean more towards European market needs and solid civil engineering projects.

Similar to the previous two BIM systems, ArchiCAD interface engages smart cursors, context-sensitive operator menus, and drag-over operation hints. Designers' sketches in the model translate into records catalogued in the document layouts, which update as the model changes. As such, the user can easily and freely merge drawing details, different model sections, and 3D images into different layouts. The user's experience of drawing via ArchiCAD mirrors the non-bidirectional process of generating reports. ArchiCAD incorporates a large collection of predefined parametric objects, including modelling capabilities for site planning and space planning. The customised parametric objects through the Geometric Description Language (GDL) are supported in ArchiCAD. The built-in object libraries provide extensive resources, such as metals; precast concrete; wood; masonry; thermal and moisture protection; heating, ventilation, and air conditioning (HVAC); plumbing; and electrical for modelling users.

The well-developed system of ArchiCAD allows multiple accesses to different domains. Internally, the home page of ArchiCAD website provides meticulous instructions on how to conduct Industry Foundation Classes (IFC) exchanges. To compensate for its shortcomings with reference to bidirectional exchange, ArchiCAD also supports a direct link to the external toolbox and other instruments, such as SketchUp and Express. Users can apply the ArchiCAD system to almost all phases of a project except fabrication. In addition, ArchiCAD has some trivial limitations concerning its modelling capabilities, as the memory inside the system is not always sufficient to handle large projects.

2.2.4 Dassault Systèmes and CATIA

Dassault Systèmes (DS) is a multinational software company that develops 3D design, 3D digital mock-ups, and product lifecycle management (PLM) software. Its BIM roots trace back to the production and manufacturing industries. DS officially commenced in 1981, but its history dates back even earlier to a group of engineers specialising in aircraft design. Whilst developing their own 3D CAD software (i.e., CATIA), the engineers broadened their scope to include automotive and AEC sector components.

CATIA was the PLM solution for 3D collaborative creation. CATIA addresses the product development process, from early product concept specification to product in service. Similar to Autodesk, DS also initiated a series of company acquisitions to integrate new products. For example, DELMIA, the company's PLM digital manufacturing software, allows manufacturers to virtually define, plan, create, monitor, and control all production processes, from early process planning and assembly simulation to complete definition of the production facility and equipment. ENOVIA furnishes a framework for collaborative management. SIMULIA is for virtual testing and simulation, automating standard simulation processes. It can be deployed across an organisation, distributing workload across computing resources and managing the simulation results to improve collective decision-making.

It is widely accepted that the AEC industry differs from its manufacturing counterpart in the sense that works are organised as projects instead of production. DS is designed to meet companies' needs for realistic product and process simulation software and to make lifelike project mock-ups more readily accessible through integration and collaboration.

2.2.5 Tekla

Tekla software offers a range of programs for design and detailing, project team review, and communication. Unlike Autodesk or DS, the family comprises several software packages, which work together but individually support clearly defined functions. For example, Tekla Structures is 3D BIM software used in the AEC industry for steel and concrete detailing, precast, and cast in situ. The software permits users to develop and control 3D structural models in different materials (e.g., concrete or steel), guiding them through the process from concept to fabrication. Tekla Structural Designer is a software solution devised for the analysis and design of buildings and thus favoured by structural engineers. Tekla Tedds is an application for automating repetitive structural and civil calculations. The software is used in engineering to create outputs, such as calculations, sketches, and notes. Tekla BIMsight is a software application for BIM-based construction project collaboration. It can import models from other BIM applications using the IFC format. With Tekla BIMsight, users can perform spatial coordination (i.e., clash or conflict checking) to avoid design and constructability issues and communicate with others in their construction project by sharing models and comments.

2.2.6 Glodon

Glodon Software is a China-based software company focused on BIM software development relating to the life cycle of a construction project. Founded in 1998 and listed on the Shenzhen small- and medium-sized enterprises (SME) Board in 2010, Glodon's products include construction engineering cost management software and project cost management software.

Glodon rebranded its flagship BIM solution as Cubicost, which aims to provide the construction industry with a more precise and expedient BIM-integrated solution through its four major products, namely, TAS (take-off for architecture and structure), TRB (take-off for rebar), TME (take-off for mechanical), and TBQ (cost estimating and creation of BoQ). One can discern that Glodon targets BIM-based cost management, which closely relates to the theme of this book. China has a relatively short history of using BoQ. First piloted in 2000 in the regions of Guangdong, Jilin, and Tianjin and three years later adopted into the National Standard (Guo Biao, GB) of BoQ for Building Projects (GB50500 – 2003), BoQ has circulated countrywide. Nevertheless, the Standard, similar to others prevailing in other economies, has its own peculiarities, which necessitates the development of BIM-based cost management solutions suitable for China. It is against this backdrop, together with China's high-speed economic growth, that Glodon is also developing rapidly. According to Flannery (2015) of *Forbes*,

> Glodon, an IT platform provider that serves design institutes and property-related businesses, had revenue of 1.8 billion yuan, or US$284 million, in 2014, an increase of 26% from a year earlier. Net profit was 596 million Yuan or US$96 million up 22%.

Based on its success in its indigenous market (i.e., China), Glodon has in recent years started to establish an international presence. Overall, it implements a focus strategy on niche markets (Porter, 1980, 1985) (i.e., BIM-based cost management) to develop its BIM businesses.

2.2.7 RIB *i*TWO

RIB is a German-based public listed company focused on construction-related software and technologies. Its flagship product, iTWO, integrates the virtual and physical worlds of a construction project. According to RIB, with iTWO, one can build a construction project virtually from beginning to end before any physical construction.

Figure 2.1 shows how iTWO concentrates on project cost management, although it can serve early design, bidding and tendering, project management, finance control, and operation. iTWO can provide virtual cost simulation from design to operation, with the whole construction process fully visualised and coordinated on screen prior to physical construction. iTWO's end-to-end virtual construction platform constitutes a 5D interface (i.e., the three given geometric dimensions plus time and cost dimensions). Curiously, RIB rarely cites BIM in its production introduction. However, in practice, iTWO accepts BIM from the designer as a point of departure for ensuing cost management functions.

RIB adopts an internationalisation strategy. Headquartered in Stuttgart, Germany, RIB has been listed on the Frankfurt Stock Exchange since 2011. It established 5D Labs in Guangzhou, Shanghai, Sydney, Copenhagen, and Madrid. RIB collaborates

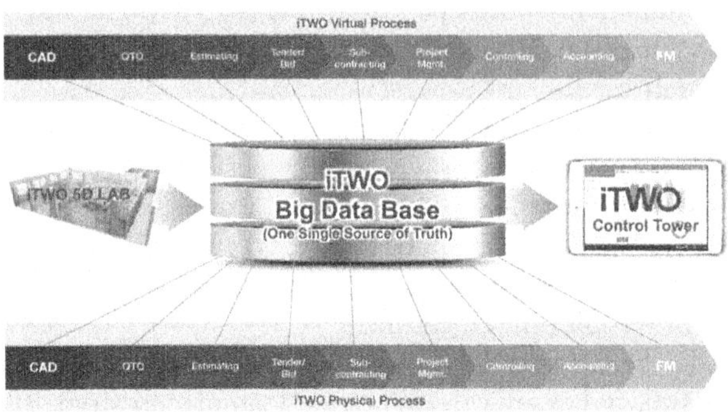

Figure 2.1 Illustration of integrating the physical and virtual world using RIB iTWO. (Permission obtained from RIB.)

closely with research institutions and universities, such as Stanford University, Georgia Institute of Technology, and the University of Hong Kong, in order to cultivate knowledge and business strategies relating to BIM. In 2018, RIB collaborates with Microsoft to embark on a joint project called MTWO. MTWO is a cloud solution that is specifically tailored to the construction and real estate industries, offering optimised cloud performance for BIM, optimised data management of construction projects on Azure virtual machines, and creating new artificial intelligence (AI) solutions to support the work of construction and real estate professionals.

To summarise, many software packages can be called BIM software and be developed to target different professions, such as architecture, structural, MEP engineering, project management, cost management, and facilities management (FM) involved in a construction project. Although most claim to offer a complete life cycle platform for AEC projects, they thrust in a specific area, and thus different professions retain different preferences. In the authors' opinion, these companies unnecessarily compete against one another. Rather, they should seek to supplement their competitors' inadequacies and mutually benefit from the enormous and global BIM demand.

2.3 Expandable BIM tools and platforms

Mainstream BIM software, as mentioned previously, mainly focuses on the architectural model and the structural model of construction. BIM can be expanded to structure, quantity estimation, fire protection systems, electrical design, energy design, mechanical systems, plan and scheduling, and so on. These functionalities usually depend on API, extensions, and value-added software packages available through mainstream BIM software or sometimes third-party software that can recognise the file format of a given BIM model. According to its main purposes, these tools and platforms divide into four categories: sustainability, MEP, construction, and FM.

2.3.1 *On sustainability*

Autodesk Revit appropriates sustainability-related extensions from the Autodesk product pool. Simulation reports span from energy to daylight and shadow analysis. Programs such as the Autodesk Green Building Studio and Energy Analysis are the most widely used.

Bentley AECOsim Energy Simulator simulates and analyses building mechanical systems, environmental conditions, and energy performances. Regarding 2D/3D energy models, documents and reports include the analysis of peak loads, annual energy calculations, energy consumptions, carbon emissions, and fuel costs. The report formats are ASHRAE Standard 90.1 compliant and Leadership in Energy and Environment Design (LEED) certified.

GRAPHISOFT EcoDesigner STAR is an ArchiCAD extension, which provides full environmental and energy modelling and simulation from BIM workflow. This extension is currently available in Australia, Brazil, Canada, Denmark, Estonia, Finland, Hungary, Lithuania, the Netherlands, Slovenia, South Africa, Sweden, and the UK.

2.3.2 On mechanical, electrical, and plumbing (MEP)

Autodesk Fabrication family (*CADmep*, *ESTmep*, and *CAMduct*) utilise manufacturer-specific content to generate interoperable solutions for better estimates and create more accurate detailed models.

Bentley Hevacomp Mechanical Designer accurately analyses building load and energy calculations using a straightforward and robust calculation methodology. The reports are based on an extensive database and comply with UK building regulations.

DesignBuilder provides advanced modelling tools in an easy-to-use interface. The results comply with LEED and ASHRAE 90.1 Performance Rating Method.

2.3.3 On construction

Autodesk Navisworks suite (*Manage* and *Simulate*) manages 4D BIM in real-time and assesses the model through convenient tools like redlining, comments, and measurements. Navisworks suite itself also has extensions that enhance 4D simulation, photorealistic rendering, interference detection, and PDF-like publishing.

Bentley ConstructSim Planner manages and distributes engineering, construction, and installation work packages by leveraging the most up-to-date project information, construction progress, and material status. ConstructSim Planner can visualise the engineering data and the 4D construction schedule.

Vico Office Suite (*LBS Manager*, *Schedule Planner*, *Production Controller*, *4D Manager*, *Cost Planner*, and *Cost Explorer*) amalgamates the 3D model, the 4D schedule, and the 5D estimate of BIM. Each take-off item has a special assembly for material, labour, and equipment. Location data help to arrange the items for an optimal schedule. The envisaged result is a cost- and resource-loaded schedule. In Vico Office, a design change in the model is immediately gaged with a new schedule and new estimate because of the tight integration.

2.3.4 On facilities management (FM)

Autodesk BIM 360 Field (together with *Revit* and *Navisworks*) is management software facilitating quality, safety, and commissioning for construction and capital projects. It primarily enhances mobile applications for FM.

ARCHIBUS combines BIM, Geographic Information Systems (GIS), and mobile devices together to improve asset, risk, and FM. Tens of third-party add-ons can further enhance the operability and interoperability.

FM: Interact is a software suite consisting of user-friendly web-based FM tools. This suite aims to improve space utilisation, occupancy, assets, moves, maintenance, leases, and property management. Beside of the technology of BIM, Activity Based Workplaces is also extensively supported in FM:Interact.

IMAGINiT Clarity Owner Data Portal assimilates with Revit model, ARCHIBUS, and Bentley ProjectWise data and exports report complying with the COBie format.

VintoCON ArchiFM (for *Graphisoft ArchiCAD*) also integrates BIM, GIS, and mobile devices together for asset planning, maintenance, and reporting services.

In summary, over the past decade, the construction industry in all major real estate intense economies has to some extent adopted BIM technology. However, the functionalities of BIM are yet to fully implement the tools and platforms listed above into day-to-day practices. One of the major reasons for this reluctance centres on chief stakeholders' diverging expertise and thus comfort using the available software. However, increasingly seamless collaboration amongst partners via integrated project delivery (IPD), and perhaps the integration of BIM functionalities, suggests a probable rise in BIM application.

2.4 Information in BIM

The keyword of BIM is 'information'. Increasingly, BIM is mystified as a panacea to the problems of the construction industry. Actually, without 'information', BIM cannot perform anything. BIM contains both geometric and non-geometric information about a project (Pratt, 2004). Geometric information includes size, volume, shape, and spatial relationships, whilst non-geometric information includes the type of individual construction component, material specifications, construction schedule, and cost. Schlueter and Thesseling (2009) suggested another taxonomy of BIM information to include geometric, semantic, and topological information. Geometric information directly relates to the building form in three dimensions; semantic information describes the properties of components (i.e., more advanced rule and function information); and topological information captures the dependencies of components. In contrast, Xue et al. (2018a) argued that there are generally two types of semantic information contained in the BIM. Semantic information of individual construction components includes geometric information, such as size, position, shape, and textures, and non-geometric information, such as type, material specifications, and meanings of functions. Then there is semantic information of relationships between components, such as dependency, topology, and joint information. Regardless of classification, semantics are invariably important BIM information.

2.4.1 Semantic information in BIM

The semantics of BIM help to interpret detailed aspects of BIM implementation, especially the aspects not visually discernible through a 3D computer presentation. Semantics includes all information concerning scopes, requirements, design (e.g., geometric, mechanical, thermal, and material properties), production, schedule and plan, maintenance, altering, and demolition. In other words, semantics explain what a BIM component was, is, and will be. Like the semantics in linguistics and philosophy, the semantics of BIM comprise the meanings and relationships of BIM elements (e.g., materials such as wood, components such as doors, and subsystems such as vertical conveying systems).

Meanings, such as the purpose of the function, geometry, and timestamps in the life cycle, of a component, explicitly describe what it is or will be. The meaning of an external wall, for example, can typically be recorded as a list of its properties within a BIM software, as shown in Figure 2.2. BIM software usually

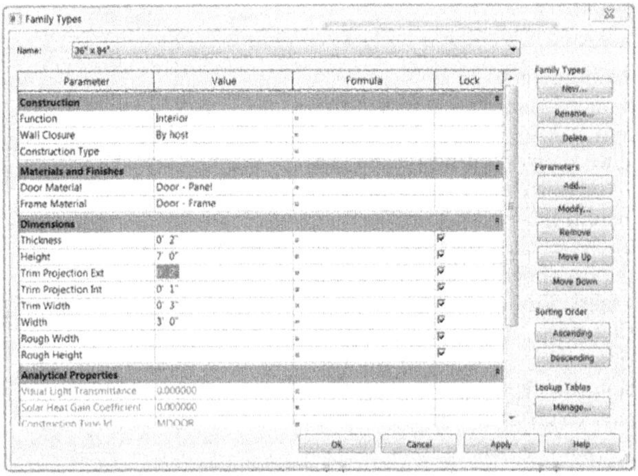

Figure 2.2 Example of property list in Revit.

has a limited set property list, but users can input additional properties for specific purposes as needed. Table 2.1 shows a list of property examples that can facilitate in simulation and project conduction.

Revit, ArchiCAD, and other prevailing BIM software programs mainly cover the spatial relationships of BIM, such as joining, grouping, and containing. Other relationships such as dependency in schedule and functional relationships in service are not included in the modelling phase in most BIM software. The user needs to manage these types of relationships on his/her own. In fact, there can be many types of relationships between BIM elements, as listed in Table 2.2.

2.4.2 *Standards relating to BIM semantics*

There are several prevailing standards in relation to BIM semantics. *Uniclass2015* organises and classifies the meanings and relationships throughout all phases of the design and construction process. *IFC*, also ISO16739 standard now, is a data format for exchanging information between different software systems.

COBie (Construction Operations Building Information Exchange) is a spreadsheet data format above the IFC standard. COBie can support operations, maintenance, and FM once in service. The delivering information, such as preventive maintenance schedules, equipment lists, spare parts lists, and warranties, instead of geometry, is focused in COBie.

NBS BIM Toolkit can provide work plans to help define roles and responsibilities for preparing information. The toolkit can also verify correctly classified objects and required data in the model.

2.4.3 *The importance of BIM semantics*

The semantics of BIM, which contains all actual values and relationships, establishes the essential information foundation for BIM analysis. Various BIM-enabled

Table 2.1 Sample of properties relating to an external wall

Property	*Type*	*Remarks*
Shop submittal parameters		
Date – issued for construction	Date time	
Date – permitted	Date time	
Date – received for shop detailing	Date time	
Date – detailing submitted for engineer of record (EOR) review\out for approval (OFA)	Date time	
Date – final erection drawings approved for fabrication	Date time	
Date – fabrication start	Date time	
Date – fabrication end	Date time	
Date – fabrication shipped	Date time	
Date – fabrication received	Date time	
Date – erection	Date time	
Date – inspected	Date time	
Wind load capacity (pressure)		MPH/pounds per square foot
Wind load capacity (drag)		MPH/pounds per square foot
Thermal resistance		R-value (h·ft2·°F/Btu)
Condensation resistance		Options: [yes, no, class]
Water resistance		Options: [yes, no, class]
Air infiltration		Options: [yes, no, class]
Fire rating		Options: [UL label – A, B, C, D, E, S]
Impact resistance (hail)		Options: [yes, no, class]
UV resistance		Options: [yes, no, class]

Table 2.2 List of possible relationships between elements of BIM

Aborts	Contains	Follows	Maximises	Regulates
Adopts	Controls	Functions as	Measures	Rejects
Aggregates	Coordinates	Gathers	Merges	Replaces
Allows	Decreases	Generates	Minimizes	Requires
Appends	Delimits	Guides	Monitors	Reviews
Arranges	Delivers	Has part	Notifies	Revises
Assembles	Demolishes	Has resource	Observes	Runs
Assesses	Demonstrates	Identifies	Operates	Samples
Builds	Deselects	Ignores	Owns	Selects
Captures	Detects	Implements	Participates in	Shares
Causes	Determines	Improves	Performs	Simulates
Certifies	Describes	Incentivizes	Plans	Sizes
Checks	Develops	Increases	Populates	Starts
Chooses	Divides	Informs	Prepares	Stops
Classifies	Documents	Initiates	Prescribes	Supplies
Completes	Draws	Integrates	Prioritises	Surveys
Collaborates	Empowers	Interchanges	Procures	Tests
Collates	Enforces	Involves	Produces	Tracks
Collects	Engages	Joins	Proves	Trains
Commissions	Establishes	Knows	Provides	Transmits
Communicates	Estimates	Leads	Pulls	Updates
Conducts	Exchanges	Links to	Pushes	Uses
Confirms	Explodes	Locates	Qualifies	Validates
Constructs	Extracts	Maintains	Quantifies	Visualises
Consults	Fabricates	Makes	Receives	Warns
Contacts	Facilitates	Manages	Recommends	Writes

analyses, such as clash detection, ventilation, daylighting simulation, buildability checking, and cost management as the focus of this book, all require detailed semantics in each model. Consider simulating the thermal performance of a house, the thermal properties and the topological relationships of all key elements of the house must be presented in the model so that the simulation can be performed.

Another important BIM aspect, interoperability, relies on the semantics of BIM as well. Different BIM software solutions have different native properties and file formats. For example, the same door often has different identifiers in the architectural, the manufactural, and the FM models. IFC specification is one of the best formats for information exchange and interoperability. The data in IFC format provide interoperability in type level between IFC compatible BIM software. Working with BIM at a level 2 minimum helps to resolve other issues of interoperability. Notably, the UK government mandated in April 2016 the application of BIM level 2 for all centrally procured government contracts.

2.5 Level of Development (LoD)

LoD is an important factor related to BIM model development in terms of the information that they should contain (Solihin and Eastman, 2015). The LoD indicates the geometric and other information contained in the BIM model and the degree to which stakeholders can rely on the contained information (BIMForum, 2017). It is thus similar to a 'common language' for BIM users to understand the information in BIM more specifically. The acronym LoD can stand for either *level of detail* or *level of development*, sometimes used interchangeably in BIM applications despite their meanings differing. Throughout the project life cycle, the BIM model can be updated, and its LoD increased as information becomes available.

2.5.1 Level of detail

James H. Clark, a famous American computer scientist, first introduced the concept of level of detail in 1976. Clark (1976) proposed a visual hierarchy for viewing the 3D model. In this hierarchy, the 3D model should become more detailed as it approaches the viewpoint of the viewer and vice versa. By actively adjusting the number of primitives (e.g., polygons) of a 3D model based on the distance of the viewpoint, this hierarchy aims to reduce the workload of the graphics-processing unit (GPU) so as to increase the processing efficiency of 3D model software.

Compared with distance-based perspective in computer graphics, when the level of detail is introduced to BIM, its definition has been modified as a feature of a certain model object. The level of detail is defined as the quantity of the information carried by each model. This definition relies on an indispensable assumption that all the provided information is tightly correlated with the project. Therefore, the level of detail of the evaluated model is valid, and the information from that model can be certainly relied upon (McPhee, 2013). Various academic institutions have endeavoured to initiate the classification and interpretations of the level of detail, whilst a universally accepted version has yet to be agreed. Table 2.3 presents two examples, which were developed by the AEC (UK) Committee (2012) and Computer Integrated Construction Research Group (CICRG) (2010), respectively.

2.5.2 Level of development

In this book, LoD refers to the level of development if not otherwise stated. Compared with the level of detail, the LoD is not necessarily a measure of the amount of information but should indicate the amount of information usable and reliable (McPhee, 2013). The most well-known LoD schema was developed by the American Institute of Architects (AIA) in its G202-2013 Project BIM Protocol Form (AIA, 2013a). The AIA's LoD schema provides a systematic way of indicating the extent of reliable information of individual objects in BIM, but the provided

Table 2.3 Examples of level of detail

AEC (UK) Committee		*CIC research group*	
LoD	*Interpretations*	*LoD*	*Interpretations*
G0 – schematic	Symbolic placeholder representing an object that may not be to scale or have any dimensional values, acutely relevant to electrical symbols that cannot be 3D object.	C	Schematic size and location
G1 – concept	Simple placeholder with absolute minimum level detail identifiable, (e.g., as any type of wall). Superficial dimensional representation. Created from consistent material either 'Concept-White' or 'Concept-Glazing'	B	General size and location, including parameter data
G2 – defined	Contains relevant metadata and technical information and is sufficiently modelled to identify type and component materials. Typically contains level of 2D detail suitable for the 'preferred' scale. Sufficient for most projects.	B (same as G1)	General size and location, including parameter data (same as G1)
G3 – rendered	Identical to the G2 version if scheduled or interrogated by annotation. Differs only in 3D representation. Used only when a 3D view at a sufficient scale deems the detail necessary due to the object's proximity to the camera. Components may appear more than once in the library with different grades. Naming must reflect this.	A	Accurate size and location, including materials and object parameters

definitions are relatively brief and narrative. Therefore, the industry would benefit more from the LoD Specification published by BIMForum (2017) that further expands upon the AIA's LoD schema. The LoD Specification provides a detailed reference to help improve the quality of communication amongst stakeholders about the information in the model. It addresses LoD100-400 of the AIA's LoD schema and adds a new level, that is, LoD350, between LoD300 and LoD400. However, the LoD Specification does not consider LoD500 since LoD500 relates to field verification but does not necessarily indicate that it contains a higher level of information. Table 2.4 presents one example of LoD mentioned in the LoD Specification and shows relevant cost estimating tasks of different LoDs.

Table 2.4 Examples of LoD[a]

Model	*LoD*	*Interpretations*	*Cost estimating*
	100	• Solid mass model representing overall volume; or schematic wall elements that are not distinguishable by type or material.	The model object may be used to build a cost estimate based on current area or similar conceptual estimating techniques (e.g., square feet of area, volume).
	200	• Generic wall objects separated by type of material (e.g., brick wall vs. terracotta). • Approximate thickness of layer represented by a single assembly. • Layouts and locations still flexible.	The model object may be used to develop cost estimates based on the assumed data provided and quantitative assessing techniques (e.g., volume and quantity of elements or type of system selected).
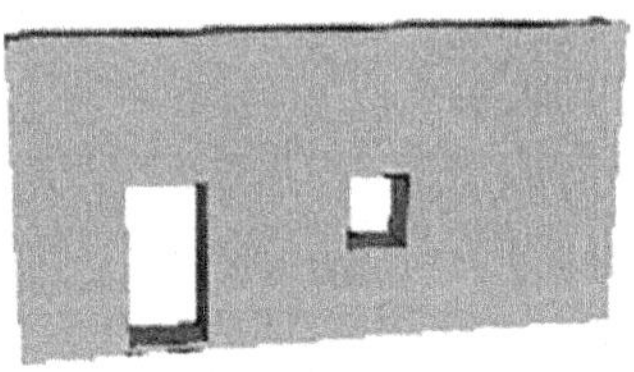	300	• Specific wall modelled to actual dimensions. • Penetrations are modelled to nominal dimensions for major wall openings (e.g., windows, doors, and large mechanical elements). • Contains shear panels.	The model object may be used to develop cost estimates suitable for procurement based on the specific data provided.

(Continued)

Model	*LoD*	*Interpretations*	*Cost estimating*
	350	• Wood framing is developed with sufficient elements to support detailed interface coordination with other systems such as MEP. • All penetrations are modelled at actual rough-opening dimensions. • Openings modelled with support framing around openings. • Elements in red are critical wall support elements that cannot be easily cut for coordination of MEP opening through the walls. • Infill wood framing modelling may be omitted at this LoD. • Cladding and sheathing are not shown for clarity in this image.	The model object contains more details for developing a more accurate cost estimation.
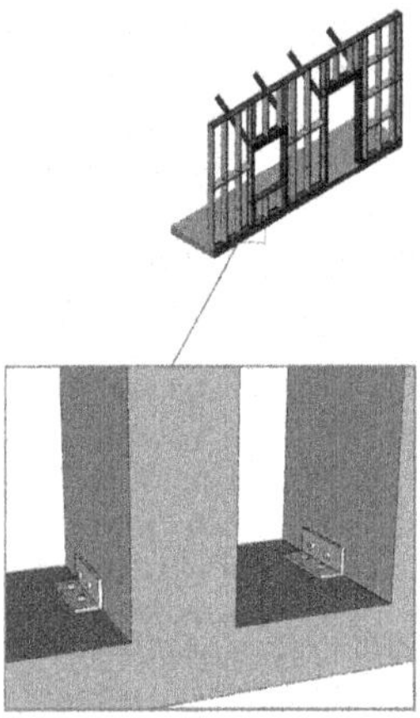	400	• Wood framing is developed with sufficient elements to support the building of the wood framing system. • Connection content is development in the wall elements. This includes but is not limited to fasteners, anchor rods, and other related hardware. • Cladding and sheathing are not shown for clarity in this image.	Costs are based on the actual cost of the model object at buyout.

[a] Models and interpretations are adapted from BIMForum (2017).

2.6 BIM standards

Many industries operate based on a set of standards, but construction is an industry that particularly relies on standards to govern its procedures, activities, and deliverables. *Standards*, as defined in the Merriam-Webster Dictionary, are principles devised and established 'by authority as a rule for the measure of quantity, extent, value, or quality'. To facilitate the implementation of BIM in the AEC industry, governments, institutes, and organisations throughout the world have published a number of BIM standards and guidelines. These standards and guidelines written for various professions in the AEC industry cover different aspects, including effective model building, digital file exchange, data compatibility, and so on. The above LoD can also be perceived as a standard to provide an agreed rule to define the information in BIM. By following these standards and guidelines, all participants can enjoy a clearer understanding of their roles in a BIM-based project.

Entering 'BIM standards and guidelines' in a search engine returns about 350,000 results. Such phenomenon illustrates a near crusade to standardise BIM implementation, and at the same time the near impossibility of discussing all BIM standards and guidelines within a single book chapter. Therefore, the authors present only those that are representative and prominent.

The International Organisation for Standardisation (ISO) publish *ISO 29481-1:2016 Building information models—Information delivery manual—Part 1: Methodology and format* in 2016. The intention of this manual centres on enabling interoperability between various software applications during all stages of a construction project's life cycle. The manual's methodology maps business processes with the information required by these processes. Another fundamental set of BIM procedures, the *International BIM Implementation Guide* published by the RICS, is designed for all types of projects regardless of size and complexity. Its purpose, to formulate best practices for BIM implementation, helps to transform and complement current practices for better project performance.

In the US, the *National BIM Standard-United States* is the most widely adopted BIM standard, which stipulates consensus-based standards by referencing existing standards, documenting information exchanges, and delivering best business practices for all building and site types. The *BIM Guide Series* published by the US General Services Administration (GSA) also provides guidance and conditions to ensure GSA projects utilise BIM efficiently with value added (GSA, 2007).

In the UK, *AEC (UK) BIM Technology Protocol* has been widely adopted by the British AEC industry since its 2015 publication. It builds on the guidelines and frameworks defined by other UK standards documents, such as the *PAS 1192-2 Specification for information management for the capital/delivery phase of construction projects using building information modelling.* This protocol aims to maximise efficiency by adopting a coordinated approach to working towards the UK government set levels of BIM maturity. A BIM maturity describes levels of maturity with regard to the ability of the construction supply chain to operate and exchange information (Liang et al., 2016). The widely cited one is the Bew-Richards maturity model (https://goo.gl/eT1usc). The Government 2011 Construction Strategy requires that, Government will require fully collaborative 3D BIM (with all project and asset information, documentation and data being electronic) as a minimum by 2016. The *Technology Protocol* also seeks to define best practices that guarantee information exchange in a collaborative environment across a project life cycle.

In Australia, the NATSPEC published the *National BIM Guide* in 2011. This guide is an important reference document that assists clients, consultants, and stakeholders to clarify their BIM requirement in a nationally consistent manner by restricting all editing to a Project BIM Brief. It also helps to define roles and responsibilities, collaboration procedures, software, and digital deliverables. For other BIM standards and guidelines, see Table 2.5.

Table 2.5 BIM standards and guidelines in major countries[a]

Country	*Organisation*	*Name of standard or guideline*	*Year Issued*
US	American Institute of Architects	Project BIM Protocol	2013
	Georgia State Financing and Investment Commission	BIM Guide	2013
	State of Tennessee Office of the State Architect	BIM Requirements	2013
	Department of Veterans Affairs	The VA BIM Guide	2010
UK	AEC (UK)	AEC (UK) BIM Protocol for Revit	2012
	British Standards Institution	BSI_PAS_1192_5_2015	2015
	British Standards Institution	BSI_PAS_1192_3_2014	2014
	British Standards Institution	PAS 1192-2:2013	2013
	Construction Industry Council	BIM Protocol	2013
Australia	Australian and New Zealand Revit Standards Committee	Australian and New Zealand Revit Standards	2012
	Cooperative Research Centre for Construction Innovation	National Guidelines for Digital Modelling	2009
	Air Conditioning and Mechanical Contractors' Association of Australia	BIM-MEP^{AUS} Practices	2012
Canada	AEC (CAN)	AEC (CAN) BIM Protocol	2014
	Institute for BIM in Canada	Environmental Scan of BIM Tools and Standards	2011
Singapore	Building and Construction Authority	Singapore BIM Guide	2013
	Building and Construction Authority	BIM Essential Guide for Contractors	2013
	Building and Construction Authority	BIM Essential Guide for BIM Adoption in an Organisation	2013
New Zealand	Building and Construction Productivity Partnership	New Zealand BIM Handbook	2014
Finland	Senate Properties	Common BIM Requirements	2012
Norway	Norwegian Home Builders' Association	BIM User Manual	2012
	Statsbygg	Statsbygg BIM Manual	2013

Country	Organisation	Name of standard or guideline	Year Issued
Hong Kong	Hong Kong Housing Authority	BIM Standards Manual, BIM User Guide, BIM Library Components Design Guide, and BIM Library Components Reference	2009, 2010
	Hong Kong Construction Industry Council	Building Information Modelling Standards (Phase One)	2015
		BIM Standards on General Building Plan Submission (Phase One)	2017
		CIC BIM Standards (Phase Two)	Forthcoming

[a] All in English.

2.7 BIM libraries

More than once, we have heard from surveyors and their managers asking how exactly BIM can help QS, the cost estimate in particular. As mentioned in Chapter 1, two essential elements for cost estimation concern quantity take-off and pricing (Aibinu and Venkatesh, 2013). The object-based BIM model with in-built geometric information with sufficient LoD allows easy capture of the objects quantities (Monteiro and Martins, 2013). The accuracy of the results depends on the quality of the used objects, and the cost data are usually stored in an external database (Jrade and Alkass, 2007), but the development direction is to link cost data directly with the objects, that is, to expand the objects with price data. Following the local standard method of measurement, QS links the quantity take-off and cost data for pricing individual building components. In any case, a BIM library that contains the BIM objects with quantity information is an essential component for BIM-enabled QS.

The objects used to develop a BIM mainly derive from three sources. First, commercial BIM software such as Autodesk Revit and ArchiCAD provides a catalogue of basic objects for developing walls, floors, ceilings, and other building components in a BIM model. Second, BIM modellers can manufacture a tailored BIM object in BIM software to meet specific project requirements. Third, the required BIM objects can be downloaded from the Internet. There are currently more than 15 online BIM libraries supplying hundreds of thousands of BIM objects in various categories. Detailed below are a representative few.

The National BIM Library (https://www.nationalbimlibrary.com/), owned by the RIBA, offers BIM users free access to over 6,500 BIM objects in 150 categories. Most of the provided objects are in the formats of .ifc or .rvt. The downloaded object compresses into a zipped folder containing both the object file in the format selected by the BIM users and a text file detailing the specific parameters of the downloaded objects.

BIMobject (http://bimobject.com/en-us), Europe's largest and fastest growing digital content management system for BIM objects, currently has more than 52,000 parametric BIM objects free to professional and studying architects, engineers, and contractors. The provided BIM objects reside in various formats, including .3ds, .lcf, .dwg, .rfa, .skp, and .ifc.

SmartBIM Library (http://library.smartbim.com/) contains more than 13,000 BIM objects in 25 main categories. All provided objects are legible to Autodesk Revit only. BIM users can hierarchically search the required objects by category and type. Through drag-and-drop functionality, the downloaded objects can easily insert into a Revit project.

The Autodesk Seek (https://seek.autodesk.com/), an online platform where BIM users can search for, download, and integrate manufacturer-specific building components and associated information, furnishes over 68,000 objects in various formats, such as .dwg, .dxf, .3ds, and .skp. In addition to the 3D models, providers can publish design and product information of the objects such as specifications and descriptions.

Productspec (https://productspec.net/), New Zealand's comprehensive database of architecture, interior design, and building products, provides more than 4,800 BIM objects in 22 main categories. Most of the available objects are at LoD 300-400. BIM users will firstly select the file format that matches their BIM software and then download the object resources for free.

Although the abovementioned online libraries deliver a large number of BIM objects free of charge, it should be noted that not all objects available from the Internet are created with equal information and LoD (Weygant, 2011). The same type of objects collected from two online libraries may have different parameters. BIM modellers should check the parameters of all downloaded objects to ensure that they have the essential information for cost estimation.

BIM libraries also have close relationships with cost management. Once the BIM model is completely developed, the BIM objects contained in a BIM library will link to the cost data and then calculate the cost of individual BIM objects (Wang et al., 2016). Currently, significant progress has been achieved to enrich cost information in BIM libraries. Yet more comprehensive developments, for example, a cost data scheme, and how it will be linked to a BIM object, and how the information is organised in various BIM libraries to suit a locality, are elaborated upon in later chapters of this book.

2.8 Summary

This chapter has demystified BIM, which has been somewhat misunderstood over the past decade. BIM is a lot more than just a digital representation of physical and functional characteristics of a facility. It follows the global trend to develop a digital twin of the physical world (i.e., the cyber-physical system) so that many functions can be performed and simulated in the cyber world before it can be applied to the physical world. BIM is not an item of software, but it must be developed using BIM software packages. This chapter reviewed the prevailing

BIM software packages and explored how they can be expanded to other built environment-related aspects. The chapter highlighted that information is the key, without which a BIM cannot do anything. The semantic information is particularly critical for harnessing the value of BIM, and its LoD was elaborated upon in this chapter. Based on the resemblance between developing a BIM and assembling LEGO, the chapter further introduced the BIM object libraries and briefly touched upon how cost data can be linked with the BIM object libraries to make BIM-enabled QS possible. After providing an understanding of the technical aspects of BIM, the next chapter introduces macro topics like BIM implementation strategies along with their prospects and challenges.

3 BIM implementation strategies, prospects, and challenges

This chapter continues to introduce BIM. Unlike the previous chapter focusing on micro and technical aspects, this chapter focuses more on macro and 'soft' aspects of BIM in relation to construction cost management. Against the global trend of mandating BIM, the chapter explores what BIM offers the construction industry. The chapter presents the costs and benefits of BIM implementation with a view to providing an understanding of the sluggish acceptance of BIM in construction. The chapter then explains BIM execution plans as various institutions define them, which imparts necessary background information for Chapter 4 wherein a BIM execution plan for QS business is developed. It continues to discuss the issues relating to BIM implementation by focusing on legal and liability, risks, organisations, and many others.

3.1 From 2D drawings to 3D models to nD BIM

It becomes clear that BIM is not simply a replacement of 2D drawings or AutoCAD files. It is not even a 3D AutoCAD. As articulated in Chapter 2, BIM is 'live'. It is a "richer repository of information" (Eastman, 1999) that can be further developed and expanded to contain other information. For example, based on the 3D for length, width, and height, a fourth, time/schedule dimension has been added to form the so-called 4D BIM. 4D BIM involves time-related information being associated with different components of an information model. As the project progresses, the components in the information model (e.g., standard floors or curtain walls) and their semantics also change. Therefore, 4D is dynamics and quite important for schedule/time management of construction projects.

Cost information is normally added as the fifth dimension to form the 5D BIM. McKinsey (2016) defines 5D BIM as "a five-dimensional representation of the physical and functional characteristics of any project. It considers a project's cost and schedule in addition to the standard spatial design parameters in 3-D". With the 5D BIM, the cost management practices can be centred on it, for example, the individual cost, the total cost, the accumulative cost, and the real cost can all be calculated and visualised in curves (see Figure 3.1) and the 4D BIM.

As different types of information integrated into the BIM model, the dimension of BIM goes even higher than 5D. Currently, there are not widely accepted

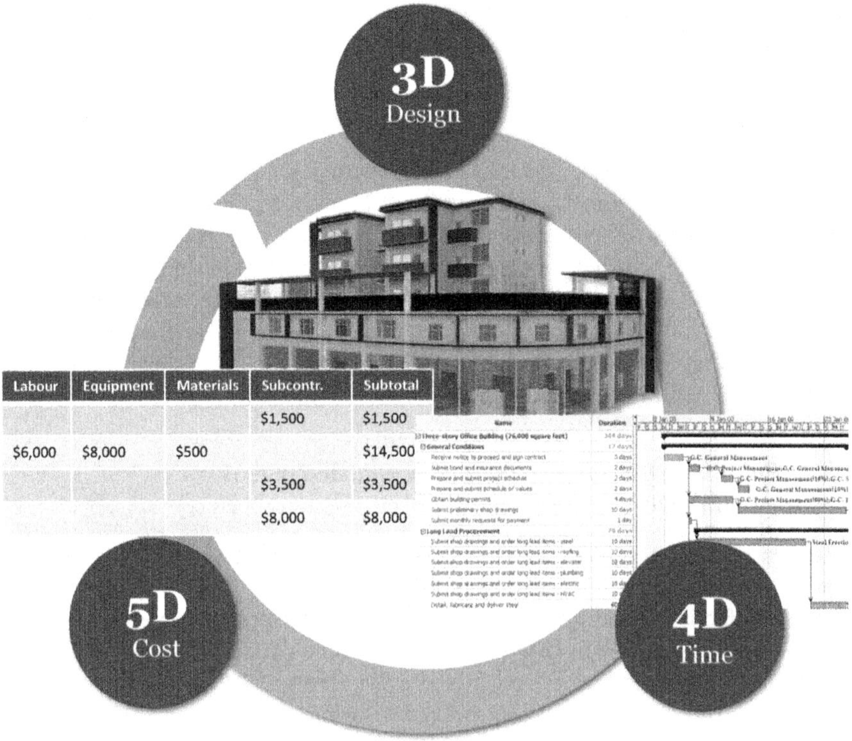

Figure 3.1 Relationship between a 3D, 4D, and 5D BIM.

definitions on BIM models with dimension higher than 5D. For example, a 6D BIM could be formed when energy information is embedded into the 5D BIM. It enables a more comprehensive and accurate energy estimates earlier in the design stage and facilitates the optimisation of design schemes to achieve a high-energy performance. It can also help to track the energy consumptions of each component. A 7D BIM is also suggested for FM. The 7D BIM is generally developed by linking operational data of building elements, such as the historical maintenance record of a pipe, to the 3D as-built BIM models handed over to the operational sphere. The 7D BIM can enable the FM sphere to quickly locate a building element in the context of operation. It can also facilitate FM department to manage operational information of the architectural, structural, mechanical, electrical, plumping, and building service elements, providing the effective and immediate access to FM information (Davtalab and Delgado, 2014).

Besides, it is suggested that other information could be also integrated into the model to achieve a higher dimension BIM. The logistic and supply chain, for example, can be integrated to the BIM as the 8th dimension, the safety information as the 9th dimension of BIM. In a sense, the idea is that with all the information integrated into a same model, BIM could provide a 'single truth platform' where different parties could get the most updated and consistent information from their collaborated parties for better decision-making.

3.2 Why is BIM in vogue?

A *Smart Market Report* by McGraw-Hill Construction (2012) found that 75% of AEC professionals in North America utilise BIM in their projects, and amongst the surveyed BIM users, 62% use BIM on more than 30% of their projects. A survey by National Building Specification (NBS) showed that nearly 74% of respondents in the UK are aware of and currently using BIM (Malleson, 2018). Countless high-profile public building offices and their developers in the UK, the US, Denmark, Finland, and Hong Kong have begun to demand BIM implementation in their projects (see Table 3.1). Why BIM is so popular? It lies in the functions that BIM can perform.

3.2.1 Enhancing productivity through virtual design and construction (VDC)

In the early stages, researchers and practitioners mainstreamed BIM as an effective way to tackle the many persisting problems of the global AEC industry, such as low productivity, poor quality, cost overrun, and excessive material waste. A widely adopted example in BIM literature is the comparison of construction and non-farm labour productivity index of the US (1964–2003). Whilst labour productivity in the AEC industry has suffered a downwards trend, shrinking to near 60% of what it was in 1964 despite the four decades (Teicholz, 2004). Although some argue that Teicholz's study is misleading because it uses labour as the sole measure of productivity (McGraw-Hill Construction, 2007) and in actuality the construction industry is creating higher, more complex structures in shorter time with higher quality (Bernstein, 2003), Teicholz's comparison nevertheless prompted the industry to rethink its problems and corresponding solutions.

BIM's 3D presentation and virtual reality (VR) simulation capabilities market BIM as more than an aesthetically stimulating design tool. BIM is a technical means devised to improve productivity. Unlike the traditional design process that expresses schemes in 2D drawings, BIM supports VDC (Kunz and Fischer, 2009) in the digital twin, allowing designers to ponder different design options and chose a best/optimal one before it is built in the physical world. As to be shown in next section, BIM can help identify potential design errors or clashes in the cyber world to avoid rework in the physical construction later. BIM supports drafting in a 3D format (i.e., what you see is what you get), if taking the aforementioned laymen's comparison between BIM and playing with LEGO. Today, many BIM component libraries are fast growing and open access (e.g., Google Warehouse, BIMobject.com, and NBS National BIM Library). These libraries boost designers' creativity and speed up turnout and, in turn, improving productivity. BIM can also maximise productivity in a number of areas, such as improving design quality, construction plan rehearsal and optimisation, and construction site management (Kaner et al., 2008; Lu et al., 2017; Zheng et al., 2017).

Table 3.1 BIM implementation roadmaps in various countries

Territory	*Implementation*	*Source*
USA	US General Services Administration (GSA) lists spatial programme BIMs as a minimum requirement for submission to the Office of Chief Architect for final concept approvals. US Army Corp of Engineers mandates BIM submission for all major building projects. US Department of Veterans Affairs calls for BIM on all construction and renovation projects over US$10M as of 2009.	HKCIC (2014), Smith (2014), Cheng and Lu (2015)
UK	Mandates the implementation of level 2 BIM on all government infrastructure projects as of 2016.	HKCIC (2014)
Netherlands	Dutch Ministry of the Interior requires BIM for all major building projects as of 2012.	Zeiss (2013)
Denmark	State clients, such as the Palaces & Properties Agency, Danish University Property Agency, and Defense Construction Service, require BIM for all projects.	Zeiss (2013), Cheng and Lu (2015)
Finland	State property services agency, Senate Properties, requires the use of BIM for all its projects as of 2007.	Zeiss (2013)
Hong Kong	The Hong Kong Housing Authority (HKHA) requires BIM for all new projects as of 2014. The government Development Bureau (DB) mandates BIM for all capital works projects with project estimates greater than HK$30M (US$3.8M) as of January 2018.	Cheng and Lu (2015)
Singapore	Mandates BIM submission for new building projects over 5,000 m^2 as of 2015.	HKCIC (2014)

3.2.2 Detecting design errors and clashes

Clash and design error detection is also one type of VDC. However, owing to its widespread implementation and immediate material implications, this function of BIM is singled out here to give special attention. Clash is a typical design error. A clash occurs when elements of separate models occupy the same space, face incompatible parameters, or appear to occur in the wrong order of the construction process's time sequence. The Association of Construction and Development (ACD, 2012) designates these hard clashes, soft clashes, and workflow clashes, respectively.

A hard clash involves two objects occupying one position. For example, a beam that the structural engineer designed is right in the path of the air conditioning units the HVAC engineer located, a column running through a wall, the slab stopping short of the wall, etc., as shown in Figure 3.2a. Hard clashes especially pervade MEP design. MEP engineers normally have to arrange all the MEP within a very confined space (i.e., the space between the main/structural ceiling and the dropped/suspended ceiling). A higher headroom can not only certainly tender better aesthetics and comfort but can also significantly increase construction costs and reduce available gross floor area (GFA), particularly in high-rise buildings. In practice, the MEP engineering outsources to different specialised subcontractors, who each compete for their respective systems' space within the given space, sometimes without knowing what has already been claimed.

A soft clash, sometimes called a clearance clash, refers to objects that demand certain spatial/geometric tolerances or buffers. An example of soft clash can be

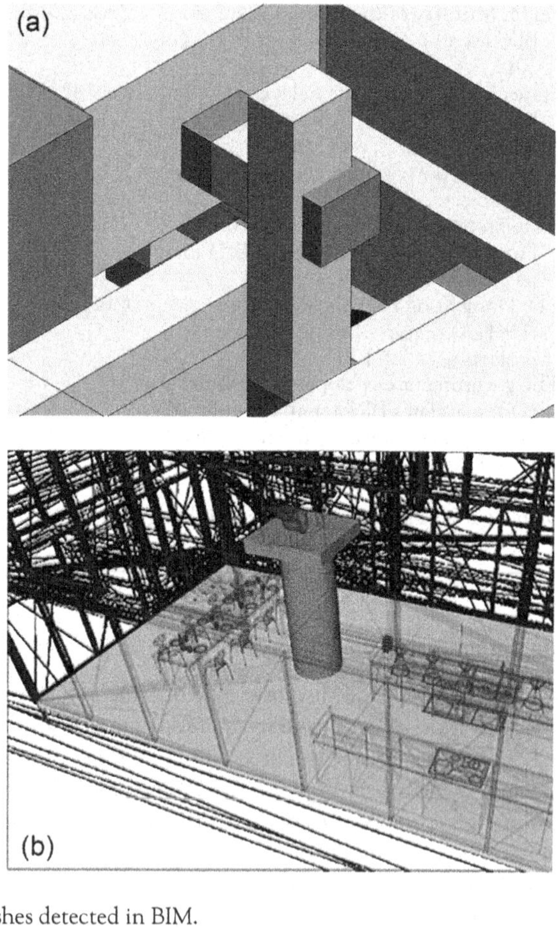

Figure 3.2 Clashes detected in BIM.

referred to Figure 3.2b. Buffer zones allow future insulation, access, maintenance, and safety. A typical example of a soft clash can be found in the case study of Barts Health NHS Trust and Royal London Hospital (Harty et al., 2010). The first floor of the hospital was set aside for imaging equipment such as ultrasound scanners, other large-scale medical devices, as well as catheterisation theatres and seminar rooms, all of which have different and competing space and MEP demands. In one of the theatres, the door of an operating theatre is too narrow to admit a large x-ray machine. In order to resolve this post-construction, the machine would have to be taken apart by the manufacturer and reassembled in the theatre at the hospital's expense. Virtual design review in BIM beforehand, however, helped to avoid this. Furthermore, given the range of soft clashes apparent in the model, the decision was made to fabricate many of the floor's interior walls modularly to allow their temporary removable during equipment change out.

Workflow clash refers to the ability of a BIM project to anticipate scheduling conflicts. Work crews, equipment/material fabrication and delivery clashes, and any timeline-specific issue constitutes a workflow clash. This can be seen in Li et al. (2009) which reported how BIM can solve workflow clashes between the construction of a standard floor and the installation of an outrigger of a high-rise building in Hong Kong.

Added expenses, costly abortive works and project delays can be unavoidable if these clashes go unidentified until the construction stage. It has been estimated that industry-wide each clash identified in the design stage saves a project about $17,000 (ACD, 2012). If a project opts for prefabrication construction technology, where steel and/or concrete components have been massively precast, errors are exponentially problematic. Clash detection should be conducted before the tendering and construction stages.

Prior to BIM, clash detection was a manual process. It involved overlaying drawings on a light table and visually catching the problems. Exclusive to senior (i.e., higher paid) designers, the process is time-consuming, insipid, and far from fail proof. In contrast, as BIM encompasses all necessary geometric and semantic information, well-developed algorithms encapsulated as plug-ins or add-ons can perform the detection spontaneously. Clash detection technology breaks down into two areas: (1) clash detection within the BIM design software and (2) isolated BIM integration tools that perform clash detection. In Revit, for example, when a slab is not connected to a wall, it will alert proactively. Depending on the tolerance level predefined as 'clash', BIM-based automated clash scans may generate a large number of 'junk clashes' that should be carefully assessed before responding to them properly.

3.2.3 *Improving interoperability*

Eastman et al. (2009) attributed the problems plaguing the construction industry, such as low productivity, poor quality, and cost overrun, to a lack of interoperability. The Association Francophone des Utilisateurs de Logiciels Libres (AFUL) interoperability working group defined interoperability as 'a characteristic of a product or system, whose interfaces are completely understood, to work with other products or systems, present or future, in either implementation or access, without any restrictions'. Wikipedia has a definition of software interoperability:

> With respect to software, the term interoperability is used to describe the capability of different programs to exchange data via a common set of exchange formats, to read and write the same file formats, and to use the same protocols.

A study commissioned by the National Institute of Standards and Technology (NIST) and led by Gallaher et al. (2004) revealed inefficient interoperability accounted for cost increases of US$6.12 per square feet for new construction and US$0.23 per square feet for operation and maintenance, resulting in a total added cost of US$15.8 billion per year. Likewise, US CII estimated that lack of interoperability led to 57% of the efforts in the design and construction industry non-value-added and thus squandered. This is a huge waste in view of the huge construction expenditure. BIM, by emphasising a shared information platform for all stakeholders of a project, can largely alleviate the interoperability problem and achieve value.

3.2.4 Reducing fragmentation and discontinuity

In addition to its direct benefits, BIM poses an indirect and likely more significant impact to construction projects by overcoming some inherent problems of the AEC industry. Fragmentation and discontinuity in the AEC industry are well known (Latham, 1994; Egan, 1998; Wolstenholme et al., 2009). "[BIM] addresses most of the trenchant criticisms made of the British way of working over decades by a succession of reports into the industry" (Saxon, 2016, p. 4). With such structural problems in procurement models, issues like risk-aversion, short-termism, silo thinking, lost information, and ineffective communication become common (Lu and Li, 2011).

BIM has been advocated as a 'soft power' for tackling these persistent obstacles rooted in the existing construction procurement and procedure. As a continuing digital platform, it can retain information (e.g., design rationale) to reduce discontinuity (Li et al., 2009). BIM can also encourage integration and collaboration (Taylor and Bernstein, 2009), particularly when it works with the IPD model (AIA, 2007a; Sive, 2009).

MacLeamy cleverly illustrated how BIM abets performance via a set of self-titled curves (i.e., MacLeamy curve, see Figure 3.3). He criticised the traditional design-bid-build (DBB) procurement system as fatally flawed:

> Today's buildings are not good enough-not because there are not enough skilled architects and contractors, but because these professionals are operating within a fatally flawed system. In the current way of working, the developer or building owner typically signs separate contracts with the architect

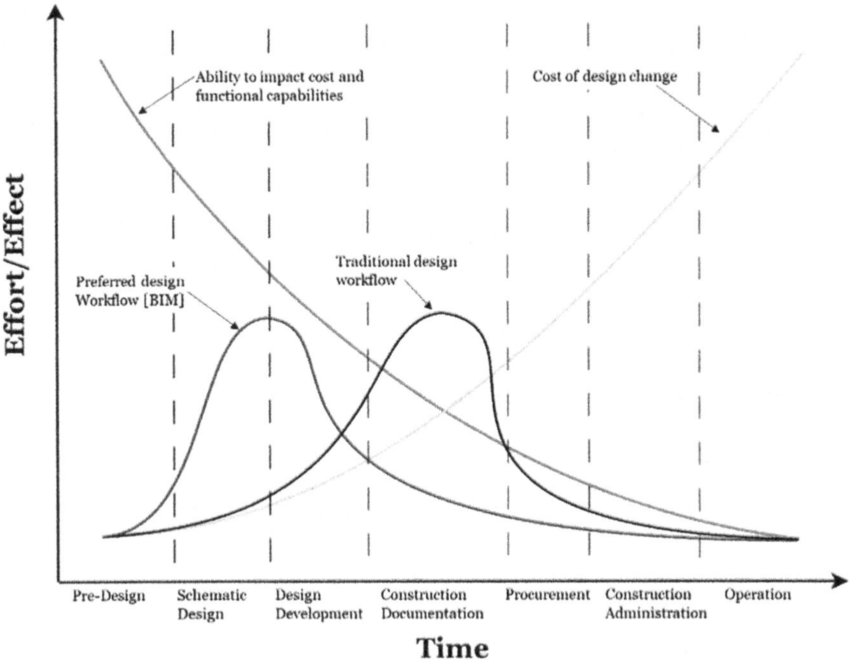

Figure 3.3 Time-effort distribution curves with or without BIM. (Adapted from MacLeamy, 2004.)

and the contractor. As a consequence, the two professionals do not always work together efficiently and, in fact, can have competing interests. Because designer and contractor do not always function as true partners, the quality of the delivered project -the building- can be compromised.

IPD is an innovative project approach that incorporates people, systems, business structures and practices into a process that collaboratively harnesses the knowledge, talents, and insights of all project stakeholders to increase project value, reduce waste, and optimise efficiency through all phases of design, fabrication, and construction (Lu et al., 2012; Jefferies and Rowlinson, 2016). It attempts to share the risks, responsibilities, and liabilities for project delivery equitably amongst the main project participants. BIM is one of the core elements around which IPD is structured (Construction Specifications Institute [CSI], 2011). Sharing similar principles, BIM and IPD naturally support one another in order to achieve better value in construction. The combination of IPD and BIM carries several benefits. For example, integrating contractors early in the planning phase largely increases the constructability of a project and thus decreases problems in the erection phase. By using integrated expertise, the design scheme should enjoy a higher quality, potentially cutting waste and energy consumption. It should be expected that IPD, together with BIM, will influence the organisational structure and enable the main principles and benefits of IPD to be implemented. Furthermore, with changing management theory and even advanced technology like BIM, the procurement model will likely evolve towards superior project performance.

3.3 BIM costs and benefits analysis

Since its advent, the diffusion of BIM has been inextricably linked with the inquiries into its costs and benefits (Lu et al., 2014). For most firms to stay afloat in such a fiercely competitive market, stakeholders have to find ways to add value and market new services to clients. Given the eminent benefits catalogued above, BIM has earned widespread recognition within the AEC industry as a state-of-the-art innovation. This has led to broad if not rapid BIM adoption amongst AEC firms for the purpose of boosting their competitiveness (McGraw-Hill Construction, 2014; NBS, 2016). Implementing BIM, however, entails both costs and barriers. Experts advise substantiating a rigorous cost-benefit analysis (CBA) of BIM implementation before deciding on a certain level of BIM implementation, rather than assume BIM will definitely generate favourable final outcomes.

3.3.1 Pertinent benefits

The beneficial impacts of BIM implementation on construction project management are manifold. As Table 3.2 shows, at the design stage, BIM helps improve design productivities, as it is useful for efficiently identifying clash and design errors (Arayici et al., 2011), and saving time and staffing (Giel and Issa, 2011). Using BIM, designers can clearly express their ideas to the client. The quality of BIM-based designs, based on 3D visualisation, is also agreed as more reliable than conventional 2D drawings. Moreover, BIM can also expand designers' business by facilitating collaborative design and offering more value-added services to clients (Liu et al., 2015).

At the construction stage, BIM can facilitate scheduling and spatial coordination before the commencement of individual working packages. In the meantime, contractors can update the BIM information and develop it into an as-built BIM that contains all detailed information of the actual situation (Chen et al., 2015). An as-built BIM is useful for the construction work by extensively supporting activities such as process control (Song et al., 2012) and of course the cost management along the construction progress.

BIM can also enable efficient FM. According to Davtalab and Delgado (2014), the two main tasks for FM generally involve preventive efforts to routinely maintain building elements, and the corrective methods concerning the failure and break down in a building operating system. BIM could assist both tasks, facilitating the locationing of building components, accessing real-time operation information, and tracking the historical record of maintenance (Becerik-Gerber et al., 2010).

Table 3.2 Pertinent BIM benefits at different project stages

Benefit	*Causal items*
Benefits at the design stage	
Design productivity	Design assessments (e.g., energy use) Higher design quality Efficient clash detection Cooperative design Efficient design changes Facilitate conceptual design Design flexibility Time saving Minimize design staffing BIM design reuse
Communication	Enhanced communication of the design intent Visualisation Information sharing
Keeping pace with market competition	Design bidding support Offer new design services Maintain working relations with the client
Benefits at the construction stage	
Construction conduction support	Reduce change orders Efficient and/or reduced shop drawings Efficient documentation Safety management Material supply management Increased construction quality Installation support/instruction Rework reduced
Preconstruction support	Construction planning and constructability assessment Clash detection Facilitate scheduling Work sequencing improved Spatial coordination
Stakeholder coordination	Improved coordination other stakeholders Reduced request for information
Keep pace with market competition	Efficient construction tendering Maintain relations with the client Client satisfaction
Benefits at the FM stage	
Facilities management	Facilities management and maintenance Facilities management tendering

3.3.2 *Cost drivers of BIM implementation*

The two highest costs imposed on BIM users are technology costs (e.g., the price of hardware, software, updates, infrastructure for interoperability, and technical support; Qian, 2012) and the cost of additional cooperation (e.g., Woo et al., 2010; Staub-French et al., 2011; Lu et al., 2014) (see Table 3.3). Variations in the working sequence, labour training, and BIM-based decision review also denote common cost items (Li et al., 2009).

Table 3.3 Cost drivers of BIM implementation at different project stages

Design stage	*Construction stage*
Technologies	Technologies
Additional coordination	Additional coordination
Ensure the data accuracy	Developing as-built BIM
BIM-based decision review	BIM-based decision review
Adding more design details	Labour training
Labour training	Organisation cost
Early decision-making	Repurposing BIM design
Space requirements	Space requirements
CAD rework cost	BIM outsourcing
Contractual cost	Risk of using a new technology
	Contractual cost
	Capital cost

Technical issues. To effectively support construction management, all objects fabricated in BIM need to be enriched in a timely manner with both the geometric and non-geometric information of physical objects they represent. Ideally, BIM should be implemented in an as-built fashion, depicting the details of how a building has been built (Tang et al., 2010). An as-built BIM is useful for extensive managerial activities, such as process control (Song et al., 2012), FM (Teicholz, 2013), design assessment (Cidik et al., 2014), and cost management (Lee et al., 2014).

Along with apprising up-to-date project information into BIM or developing an as-built BIM, projects also face considerable costs should they solve unforeseen problems, such as incompatible technologies (Bryde et al., 2013). Project managers and other foremen also have to identify BIM objects quickly and precisely and link them to a physical project (Chen et al., 2015). Developing and promoting standards associated with interoperability issues across different BIM software and regulating BIM practices, like setting consistent names for BIM objects, remain expensive and regular hurdles. Moreover, as BIM does not generally capture the information exchanged between BIM users via different media, considerable costs can surface during registry of communication and information exchange. Last but not least concerns data security. Striving for adequate high-tech maintenance and enhancement of a BIM implementation system's security, including the data security of the server, is vital to protecting intellectual property (IP) rights.

Managerial issues. Since using BIM requires firms to manage projects collaboratively and interdependently, managerial-related issues can become more expensive. Traditional AEC processes enjoy standardised workflows and are thus better positioned to anticipate the costs allied with managing people. Designers tasked with composing 2D drawings are generally expected to draw typical parts once and then nominally modify the drawing set as need. BIM of course enables designs of far greater detail, which invents new costs for AEC firms (e.g., those associated with time and intellectual efforts, Kaner et al., 2008; labour training, Qian, 2012; and ensuring the data accuracy, McGraw-Hill Construction, 2010). Such BIM capabilities meanwhile call for more collaborative design compared to traditional manner, which means considerable time, labour, and cost of review BIM-based decisions (DPR Construction, 2010). These issues may require designers' early decision-making, which implies additional costs and risks for the designer (McGraw-Hill Construction, 2010).

Added costs can also incur in the construction phase. When building with BIM, contractors often have to repurpose a BIM model for their own requirements or even reproduce it in part or entirely (Becerik-Gerber and Rice, 2010). Updating BIM or developing an as-built BIM that informs contractors can instigate considerable labour training expenses (Becerik-Gerber and Rice, 2010). As the construction phase progresses and brings on more subcontractors, suppliers, logistics planners, etc., contractors rack up organisational costs tied to obliging these partners to work around BIM and adapt their customary workflows to BIM-aided processes (Merschbrock and Rolfsen, 2016).

3.3.3 *The importance to have a BIM business case*

The CBA studies reviewed above are only a tip of the iceberg. There were numerous studies on BIM's CBA (see Barlish and Sullivan, 2012) and more are coming. The proliferation of BIM's CBA actually reflects the industry's anxiety to find itself a business case for BIM adoption. This is understandable; if a technology initiation is to sustain in a competitive business world, it must have a genuine business case. In BIM adoption, research has shown that one of the major hurdles is the justification of the additional cost using evident benefits (Li et al., 2009). BIM implementation needs a business case. Users who are to adopt BIM need the encouragement of empirical evidence, whilst investors need to discern clear proof of its benefits in order to justify their investment of time and budget.

The prolific CBA of BIM implementation can also explain the lacklustre adoption of BIM in real-life projects. It is not uncommon that companies take up a 'wait-and-see' position or simply pay lip services to BIM adoption. Without solid evidence of BIM surplus, it is unsurprising to see that many companies are hesitating in fully engaged in BIM adoption. Following the early stage of identifying anecdotal evidence of BIM benefits, recent years have seen more studies trying to identify empirical evidence of such BIM benefits, and this trend will continue. However, the difficulties of such CBA are also obvious. First, it is difficult to disentangle BIM-related costs/benefits from the rest (e.g., other technologies or innovations). Second, it is always difficult for analysts to access quality data for the task. Last but not the least, costs and benefits mean different things to different people. It is important to consider different perspectives in conducting a BIM CBA.

3.4 BIM execution plan

The BIM execution plan is of vital importance to promote BIM implementation in real-life AEC projects. A sound BIM execution plan should cover the following aspects of information:

- Objectives, project milestones;
- Key deliverables and performance tracking methods;
- Roles and responsibilities, team organisations, relevant authorities, and approval processes;
- Collaborative processes and working procedure;
- Project implementation plan;
- Task information delivery plan (responsibility for delivery of each supplier's information);
- Master information delivery plan (when project information is to be prepared by whom and using what protocols and procedures);
- Software and versions to be used and supporting hardware requirements;
- Protocol, data formats to be used for collaboration;
- LoD at different stages;
- Information management system;
- Quality control and assurance; and
- Other agreed rules for metric system, file name conventions, tolerance value, attribute data, annotations, abbreviations and symbols, etc.

With a comprehensive BIM execution plan, all project stakeholders can better understand their roles and responsibilities and the resources necessary for BIM implementation. Moreover, the BIM execution plan provides the solid reference basis for project team members to follow and tract their progress to gain the maximum benefits of BIM implementation.

The BIM execution plan should be developed in the early stage of a project by representatives from all project stakeholders (e.g., owners, designers, contractors, engineers, major specialty contractors, facility manager, and project managers). All primary team members should champion the planning process, following a structured procedure during the early phases of a project. The following subsections briefly introduce the process of making a BIM execution plan based on the CICRG (2010).

3.4.1 *Specifying BIM objectives at each project stage*

At the project planning stage, the planning team, led by the key decision-makers from various disciplines, shall propose the overall project goals that can be achieved with the help of BIM. Common goals may include reducing project duration cost or increasing the overall project quality. These goals shall be further broken down into specific, feasible, and measurable BIM objectives at each project stage to strive to improve the successes of the planning, design, construction, and operations of the facility.

After defining the BIM goals and objective at each stage, the lead BIM coordinators for each party can develop and execute the detailed implementation processes and information exchanges. The planning team shall identify the suitable tasks that the team would like to perform using BIM. It is necessary to identify the appropriate uses of BIM by beginning with the potential end-uses of the information embedded in the model. Moreover, the planning team shall rate the capabilities of each party for each BIM use and identify additional value and risk associated with each BIM use. Finally, the planning team shall determine whether to implement each potential BIM use based on the evaluation of capabilities, value, and risk.

3.4.2 *Mapping out BIM procedure*

After identifying the goal and use of BIM, the implementation process for each BIM use shall be specified and designed. Process-mapping techniques would be used to effectively understand the overall BIM process, identify the information exchanges that will be shared amongst multiple stakeholders, and clearly define the various processes to be performed for the identified BIM uses. First, an overview map shall be developed to show how the different BIM uses will be performed. Then, detailed BIM use process maps shall be built to define the specific BIM implementation in greater detail. The overview map illustrates the relationship of BIM uses specific to the project. The overview map also contains the high-level information exchanges that occur throughout the project life cycle. Important to note, each project and company is of course unique, so there may be numerous potential methods a team could employ to achieve a particular process. This shall be based on the planning team's consensus.

3.4.3 *Defining information flows*

Defining information flows holds a lot of weight as downstream BIM uses are directly affected by what is produced by upstream use. The project team must decide who shall be authoring this information and when this information needs to be placed into the BIM. The information exchanges which occur between the project participants shall, therefore, be clearly identified. It is important for the team members, in particular the author and receiver for each information exchange transaction, to clearly understand the information content and LoD.

The key procedure for creating the information exchange requirements begins with identifying each potential information exchange from the detailed BIM use process map. Then, choosing a model element breakdown structure for the project, identifying the information requirements for each exchange (i.e., the output and input), assigning responsible parties to author the information required, and comparing input versus output content. The structured procedure incites a greater understanding of the information required for exchange throughout the entire process. The information exchanges will illustrate the model elements by discipline, LoD, and any specific attributes important to the project.

3.4.4 *Selecting BIM software, hardware, and human infrastructure*

It is also important to consider all the resources and infrastructure required to perform the specified processes as BIM still faces certain trials in practice. For each BIM use selected, the project team shall determine the personnel who will perform each use; establish a plan for adapting each BIM use based on project size, complexity, LoD, and scope; and determine which personnel will typically monitor it. The planning team shall also assess the software and hardware needs of each BIM use and compare the technical infrastructure needs to the current software and hardware. The assessment of BIM software starts with determining the software applications that can suit the specified BIM objectives. The planning team shall also reach an agreement on whether to use proprietary formats (e.g., .rvt) or open formats (e.g., IFC) in order to ensure a smooth data exchange. The assessment of BIM hardware includes but not limited to determining the hardware specifications, database storage, and network bandwidth. Necessary upgrades and purchases shall be made to ensure that the software and hardware do not limit the successful execution performance. Investing in compatible programs for all major project participants for the length of the project prevents technical complications arising in the future.

3.5 Issues concerning BIM implementation

An evolving marvel for the construction industry is that BIM is more than just a facilitating technology, but rather a game changer. Its success or failure depends heavily on many economic, legal, cultural, or organisational issues surrounding this technological development (Woudhuysen and Abley, 2004; Azhar, 2011; Bryde et al., 2013).

3.5.1 *Contractual framework for incorporating BIM*

Legal fears may be acting as a hindrance on BIM's wider implementation within the construction industry (Arensman and Ozbek, 2012). The relationships amongst clients, designers, and contractors might be changed by BIM implementation. By binding to BIM, the stakeholders may not follow the traditional life cycle process. BIM may involve more complicated and relatively new collaboration between consultants, specialist manufacturers, contractors, subcontractors, and facility managers.

During project planning, BIM use and responsibilities shall be written into the contractual agreement. Due to the substantial nature of the changes full adoption of BIM promises, it is better to draft new contract templates requiring the various relationships (i.e., employer and consultant, contractor and subcontractor, client and designer, etc.) with respect to BIM working methods. The alternative would be the incorporation of legal issues in a BIM protocol, such as a set of amendments to the main contract to make it favourable for BIM. The amendments could then be incorporated into the various agreements of the project, ensuring that BIM-related rights and obligations can be rendered in different contracts. This allows parties to adopt BIM, whilst retaining the contracts they are accustomed to. This is an important point, since in practice the legal terms of the BIM protocol may conflict with the clauses of the principal contract (e.g., a BIM protocol should require a more comprehensive IP licensing procedure than that provided under current construction contracts used in the industry).

In addition, the extent of contractual and legal issues that should be covered in the BIM legal amendment (i.e., waivers, indemnities, liability for contribution, etc.) makes it inappropriate to recommend a stand-alone amendment. When the BIM amendment specifies multiple models, the parties should point out which model takes priority in the event of conflicts of information contained in each BIM model. The parties need to agree on the format and content of each model and the required standards to be employed. These may be elaborated upon in a BIM execution plan part of the contract documents. Nevertheless, none of these is readily accepted in practice. They are still in strong debates.

3.5.2 Intellectual property (IP) rights

IP refers to creations of the mind (e.g., inventions, literary and artistic works, symbols, names, and images used in commerce). These are divisible into two categories, industrial property and copyrights. The BIM model consists of contributions from various parties. To avoid liability for infringement of third-party IP rights, the protocol should ensure that all contributors warrant that they hold the IP rights over the contributions they have made and provide an indemnity to all other parties who may use such contribution in the event of a third-party IP dispute. Besides, it may be easier for users of a BIM model to duplicate the parameterised BIM component design for high efficiency. Therefore, it is necessary to identify the IP rights of each party.

However, when two parties make a distinctive addition to a contribution, it is not possible to separate where one ends and the other begins. This justifies the application of joint authorship. It seems the better approach would be to embrace the concept of joint authorship, whilst preserving the limits of the original contributor's liability. One way to achieve this is to define in the amendment what would constitute joint authorship and recognise the right of the original author to accept or reject any addition and be saved from any liability for errors wherever an addition is made without consent. Where consent is obtained, both joint authors hold rights to the contribution, as well as bear joint and separate liability for its errors.

The nature and ownership of the final product of the BIM process, especially in terms of future uses for FM, lies at the heart this debate. The end product of the BIM process should be a BIM model and a completed project. The ownership of the completed project is not in question, just the ownership of the BIM model. Designers in traditional construction have intellectual rights to their project designs, and as a result, they are often reluctant to permit the use of the model by others. Therefore, some argue that as BIM does not change the inherent principles of design, ownership should lie with the designer (Hurtado and O'Connor, 2008). However, as several different organisations work on the BIM model in collaboration, who the real designer is not so evident. Some hold the view that ownership of the final model belongs to the client. This creates legal difficulties as the BIM model contains information from different parties and each wish to retain the IP rights to the model they have contributed to (Larson and Golden, 2008). The loss of control over the model ownership becomes a legal barrier to the implementation of BIM.

3.5.3 Model management

Whether the parties contribute to a single model or federation of models, the process demands management. The model manager must implement the parties' agreement on model access, security, transmission, archiving, transmitting, etc. This stirs up several legality questions, such as who appoints the model manager, how is the manager to be replaced by, who bears the costs, if any, arising from this role, etc. The predominant thinking in this regard rests on the employer taking charge of these issues; however in a 'Design and Build (D&B)' contract, the principal contractor may be more suited. It is recommended that parties should make a suitable choice based on their procurement methods.

The powers and responsibilities of the model manager should be thoroughly set out in the protocol, just as all stakeholders' roles should be. Whether a single person or multiple people occupy these roles, the relationship amongst the model manager, the project manager, and other team members should be plainly defined, especially if the model manager may be replaced during the course of the project. To avoid conflicts, it may be worthwhile to grant the model manager exclusive responsibility and power to issue-binding instructions on BIM-related issues.

3.5.4 Risk management and liability

Reliance on data. The main advantage of the BIM model or models lies on the information it contains. This can also prove to be its great disadvantage because if the data are unreliable, so is the model. Traditionally, project designers (e.g., architects and engineers) have managed the 'risks of reliance' by prohibiting dependence by parties other than their direct client and by creating caveats supported by the law, limiting their liability to application of the relevant standard of professional care and due diligence. The BIM process should give various parties direct access to the model, allowing them to make informed decisions and allocate resources. This potentially creates new classes of possible liability. To manage this risk, it is advisable to include in the BIM protocol a clause demanding all project participants waive the rights to consequential damages and limit liability to direct losses connected to the project. Resolving who has the responsibility to ensure the quality of contributions to the model or models and the level of reliance to be placed on such contributions is also crucial, especially if the model contains intelligent objects that may change on account of information derived from other contributors.

Cost and time issues. The BIM process could throw up costs and time issues outside the principles of time extension contained in the main agreement. The implementation of BIM technology increases the benefits of communication between designers and contractors by integrating the operating interfaces. However, it may not reduce designers' fees and maybe even increase the price of revising design schemes. The parties should allocate within the BIM protocol, which party bears accountability of delays or costs arising from errors in the transmission and use of information during the BIM process. Whilst such errors will hopefully be the exception rather than the norm, the BIM protocol should afford such eventualities. These costs and time issues lie within the parameters of ownership of the BIM process. The party 'owning' the process would be the one to determine the criteria for extension of time due to BIM errors and means for added payment if necessary. The owner's liability for non-technology-related errors downgrades when the liability for error rests with the participant responsible for the error.

Software and technology issues. The technology on which BIM is based may create specific liability issues. Blanket limitation of liability clauses protects software manufacturers, transferring the risk of BIM errors caused by the software to the manufacturers. This risk should be dealt with contractually and should be borne by the party who takes responsibility for the BIM process. For software- and technology-related issues, the parties should map out carefully the liability between consultants who chose these technologies and have expertise of them, and the employer of the BIM process. The aim should be for the consultants to bear the risks for errors in the choice of technology that could have been avoided had proper care and diligence been applied whilst the residual risk lies with the owner of the process.

Integration issues. The software to apply BIM technology varies, as do associated risks and liabilities that may occur due to incomplete transfer or missing information. The protocol should identify who is responsible for the model integration. Ensuring all parties take out the appropriate insurance to cover their involvement in the BIM process is necessary if an integrated project-wide insurance is in place. Where such coverage is not available, parties should consider obtaining traditional coverage to protect against the liability for errors or omissions in their contributions to the model.

3.5.5 Organisational issues

BIM does not operate in a vacuum. Rather, it is closely related to organisations based on which BIM can reside. As mentioned previously, AEC works are normally organised as the organisational form called project-based organisation (PBO) (Söderlund, 2011); it is an industry in which the forms of PBOs have long been taken to be the norm across a significant swathe of activity (e.g., Winch, 1989; Gann and Salter, 2000; Bresnen et al., 2004; Morris et al., 2011). The formulation and operation of a PBO is determined by the procurement models (e.g., DBB, D&B, or public private partnership) adopted in the project. Readers are probably familiar with DBB, which has been accused of the main culprit of fragmentation and discontinuity of the AEC industry, and BIM is considered promising to alleviate the problems associated with DBB. In short, to discuss BIM implementation, strategies cannot disconnect it with the organisation it resides. The prospects and challenges of organisations and BIM adoption will be further discussed in Chapter 7.

3.6 Summary

This chapter further introduced BIM, by focusing on its macro and soft aspects. BIM as a hallmark development rapidly sprawls the global construction industry since it has the potential to solve the many ingrained problems that have beset the industry, for example, low productivity, poor interoperability, poor information and knowledge management, and fragmentation and discontinuity in this industry. Running in parallel with the effort to find anecdotal evidence of these benefits is the constant inquiry into its CBAs using empirical data. Another important strategy to encourage BIM implementation is to develop a sound BIM execution plan. This has been introduced as a general strategy, whilst in Chapter 4, a specific one on QS is developed. This chapter also discussed the many issues relating to BIM implementation. It focused on legal issues, risks and liability, organisations, and many other issues. It is hoped that the first three chapters have provided sufficient background information on construction cost management and BIM to facilitate the understanding of their combination in the next chapters.

4 Adopting BIM for cost management

Following on the introduction of cost management in Chapter 1 and BIM in Chapters 2 and 3, this chapter aims to link the two. It starts with the prospects of BIM for cost management. Then, it develops a building information model for quantity surveying (QS-BIM) execution plan, which can be perceived as an extension of the general BIM execution plan described in Chapter 3, with a focus on QS. Next, the critical success factors of adopting BIM for cost management are presented. These include the demand for a QS-BIM, which aims to define how such BIM should look like when developed for cost management. By clearly demanding QS-BIM, it, hopefully, can help elevate the stance of QS profession in the overall project delivery. Certainly, successful BIM adoption for cost management is not simply to demand an informative BIM. It also relates to the compatibility of BIM with existing QS practices and software development.

4.1 Prospects of BIM for cost management

Following the RIBA's Plan of Work, Figure 4.1 illustrates how BIM can help with QS' tasks and responsibilities at various work stages. Readers may notice that the figure is an extension of Figure 1.1 by elaborating how BIM can be used in different QS tasks at different work stages. In the preparation stage, a schematic BIM can be developed, which can be used to establish a preliminary cost estimate if linked to the cost database. The BIM-facilitated preliminary cost estimate can provide sufficient information for feasibility study and provide advice on procurement methods.

In the design stage, the schematic BIM will be developed into the as-design BIM containing the consolidated design information. BIM can help develop BoQ by automatically extracting the components counts, area and volume of spaces, as well as material quantities from the design. BIM can also improve the understanding of complex building structures and systems by providing 3D visualisation of construction objects and related elements. Both functions help to provide accurate BoQ (e.g., by considering deduction of component dependency or joints) and other contractual documents for tendering.

At the construction stage, the information contained in the as-designed BIM will be updated to an as-built BIM by synchronising it with the actual construction information. The as-built BIM can help track construction process and

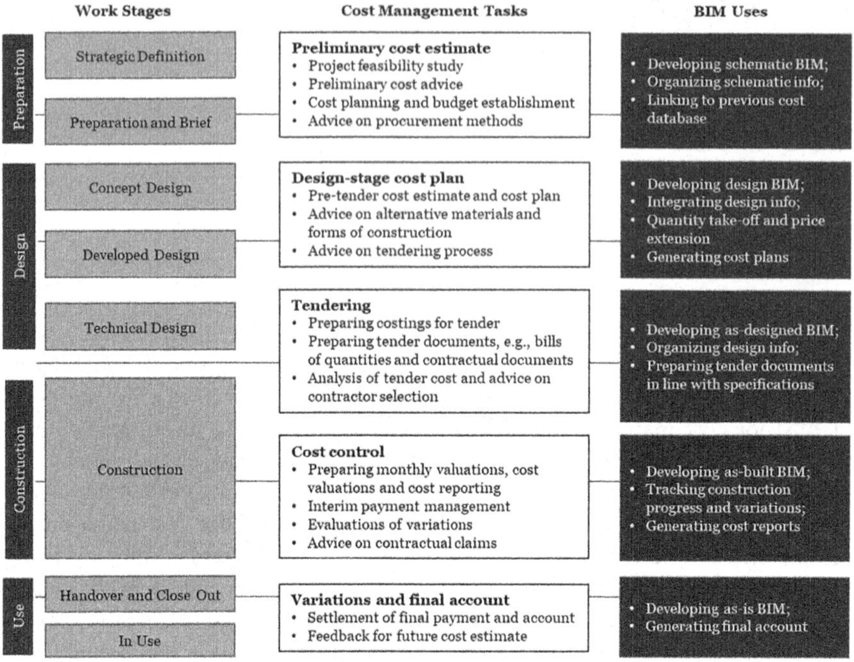

Figure 4.1 QS' changing tasks and responsibilities with BIM implementation.

variations, based on which monthly cost reports can be developed and the interim payment can be settled.

On completion of the construction work, an as-is or as-built BIM will be developed to reflect the current as-is condition of the constructed project. The as-is BIM provides the necessary information to generate and settle the final account. The as-is model can also be used for 'soft landing' (Pittard and Sell, 2016), a process to ensure a smooth ride for all concerned as buildings move from construction to occupation and use (i.e., FM).

4.2 Developing a QS-BIM execution plan

The implementation of BIM for QS practice generally involves various parties to develop, maintain, and analyse BIM for cost-related tasks. It concerns not only the technical issues of using BIM, such as setting up BIM software or hardware and defining BIM information flow therein, but also the soft, managerial, and organisational issues to truly integrate BIM into the workflow. Therefore, it is of vital importance to develop a reliable and detailed execution plan to provide guidance of BIM implementation amongst various stakeholders, specifying the BIM tasks, objectives, process, and responsibilities of individual parties, and software and hardware requirements for QS.

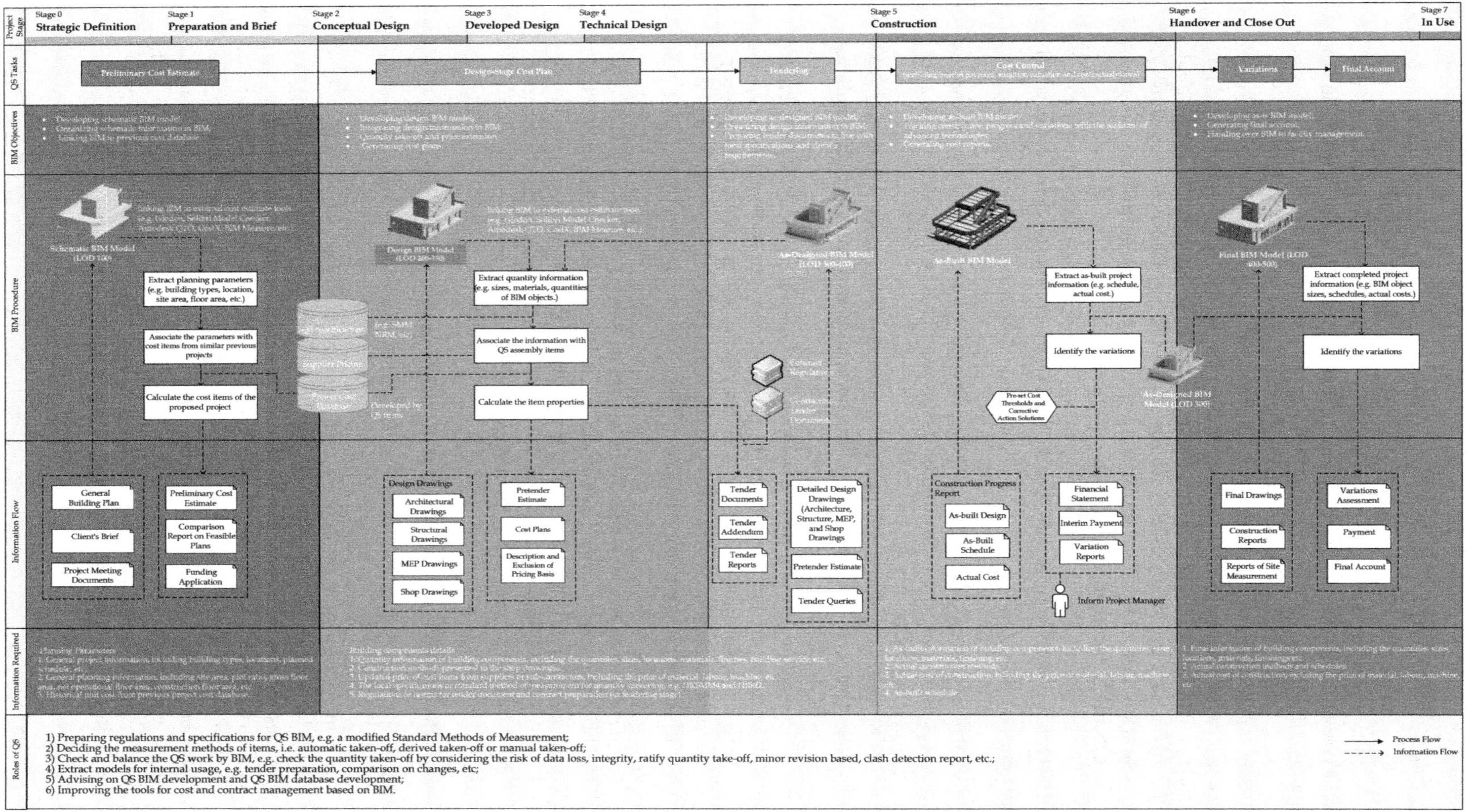

Figure 4.2 Detailed BIM execution plan for life cycle construction cost management.

As suggested by CICRG (2010), there are generally four steps to develop an execution plan, namely (1) specifying BIM objectives at each project stage, (2) mapping out the BIM procedure, (3) defining information flow therein, and (4) selecting the software, hardware, and human supports. Readers may go through Section 3.3 of this book to review these steps in detail. Following on the general procedures, this section will develop a BIM execution plan for QS practice, referred to as a 'QS-BIM execution plan' in the following sections, and provide an example for it, as shown in Figure 4.2. One size does not fit all; it should be understood that the QS-BIM execution plan is general to cover various QS application scenarios. QS companies, in real-life cases, can further tailor-make the QS-BIM execution plan for their own needs.

The proposed QS-BIM execution plan consists of four elements to implement BIM for cost management tasks during the construction project life cycle, which include:

- *BIM objectives*, which are usually proposed by representatives of various stakeholders at the beginning of the project. This is a version for cost management by QS;
- *BIM procedure*, first developed as an overview map to show how the BIM objectives could be achieved and further refined with more details specifying QS-BIM implementation process;
- *Document/information flow*, developed by choosing a model element breakdown structure and identifying the information input and output;
- *The software/hardware and responsibilities of QS*, for example, the supporting software, hardware requirements, the evolving roles of QS and collaboration with other parties in the context of BIM.

The four elements are organised in the matrix framework as shown in Figure 4.2. The horizontal axis is formed by the different work stages as proposed by RIBA's Plan of Work, and the vertical axis of the matrix is the four steps proposed by the CICRG's BIM execution plan. Readers can thus easily compare it with Figure 4.1 and Section 3.3 to grasp this QS-BIM execution plan.

4.2.1 Preliminary cost estimate process in the context of BIM

4.2.1.1 Identifying the BIM objectives

The primary aim of a preliminary cost estimate is to forecast the most likely cost of delivering a project, providing a solid basis for the subsequent cost-related activities such as setting up the budget, raising project funds, and construction contracting. At the preparation stage prior to design, only limited information is available on the project. Thus, the preliminary cost estimate relies more on the general project planning parameters, such as project size, type, and location, instead of detailed design or engineering information (Sabol, 2008). With such planning parameters, QS will then determine the estimated cost based on previous projects of a similar nature and characteristics.

Even though the estimate generated through this process can provide guidance to choose or modify different schemes, it could fluctuate significantly among measurements by different QS even on the same project. This is because QS normally rely on their own experience and expertise to extract useful planning parameters and forecast the cost based on previous projects. They usually have different perceptions on modifying the forecasted construction cost and attitudes towards construction risks. Besides, it is also difficult for QS to search for a 'most similar previous project' as the estimate reference, let alone the similar cost items from that project. This situation has triggered the need to identify an efficient way to integrate project information, extract the useful one for estimate, and quickly link the current project to the previous project database to generate a cost estimate.

BIM is expected to enable an efficient preliminary estimate and generate reliable estimation results to facilitate the subsequent cost-related tasks. Specifically, the primary objectives of using BIM should involve developing a schematic BIM model to organise the schematic design information, as well as linking the BIM model to the external database to efficiently extract cost items from the projects of similar types and characteristics. A schematic BIM model is helpful to organise the planning information, as well as visualise this information in a 3D presentation, thus enabling a better understanding of design ideas especially in the early stage of the project. Besides, when the design ideas are digitalised in an information model, it is possible to use computation power for an efficient and automatic calculation, such as linking the previous project database. In this way, preliminary cost estimating is no longer relying solely on the expertise and experience of individual QS, but the joint efforts and knowledge pooled up from previous projects.

4.2.1.2 Mapping out BIM procedures and information flow

As suggested in the BIM implementation plan (see Figure 4.2), BIM-based preliminary cost estimation begins with a schematic BIM model. The model should be provided by the architects and handed over to the client and its appointed QS for further analyses. The schematic BIM model is usually developed in simple geometries, as minimal project information is available at this stage. The model should be flexible and lightweight enough and instantly visualised to support pondering multiple alternatives and iterations. Therefore, the BIM objects in a schematic BIM model are usually in LoD 100, i.e., a generic representation with little information (see also Figure 4.3 and Section 2.5).

With the schematic BIM, the next step is to link it to cost estimate tools to further analyse the quantity information embedded in the model. The cost estimate tools will extract useful information from BIM, especially the project information from document packages, such as general building plan, client's brief, and project meeting documents. Suggesting by the HKIS (2016), the planning information may include:

- Site area,
- Plot ratio,
- Gross floor area (GFA),

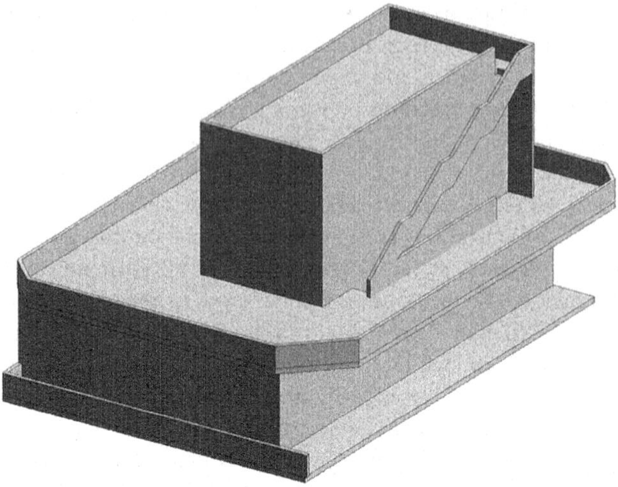

Figure 4.3 A schematic building information model.

- Net operational floor area (NOFA),
- Construction floor area (CFA),
- Guestroom mix (for hotels),
- Number of beds (for hospitals),
- Number of seats (for auditorium/theatres), and
- Number of workstations (for office fit out).

The extracted information will be subsequently associated with QS items predefined in estimating template provided by the cost estimating tools. QS are encouraged to double check the information and complete the template if the software fails to extract all the necessary information from BIM. Then, the estimating tools will further analyse the data by linking the data to the cost database of previous projects. After that, the BIM tools may also help to generate reports on preliminary cost estimate, feasibility study, and funding application.

4.2.1.3 Supporting technologies and QS' responsibilities

To achieve the aforementioned objectives, BIM software should, from a technical perspective, facilitate an efficient modelling of the schematic BIM model and be able to be linked to the external database of previous projects. Generally, there are two methods to achieve that. The first method is to develop a BIM object library, where the object information is derived from previous projects. The objects in the library do not necessarily contain comprehensive semantics. Rather, the BIM objects should be clearly defined in terms of dimension (e.g., length, width, and height), materials, cost, manufacturer, positions, orientations, and project-related information such as project type, locations, size, and duration. The semantics can be further detailed and enriched in the subsequent stages. The BIM object library should also provide an efficient search engine for designers to identify a usable BIM object quickly. In this way, a schematic BIM model could be developed quickly whilst allowing for an instant rough calculation on the preliminary construction cost.

The second method relies on software programs that link the BIM model to the cost estimate software and database of previous projects. Preliminary applications should also have the capability to export developed information to full-fledged BIM modellers (Sabol, 2008). Popular tools for cost estimate include Glodon, Solibri Model Checker, Autodesk QTO, CostX, and BIM measure. The BIM software should also provide easy-to-use functional modules to create some lightweight BIM models, so that QS can quickly create a model if the architect fails to provide one.

With BIM integrated into the QS practice, it is expected that QS could be relieved from tedious work searching the 'most similar cost items' from previous projects and calculating cost by formulas. QS can focus on developing the previous project database containing BIM and QS-related information. Besides, QS are encouraged to further check the cost estimates by BIM, further revise them based on their own expertise and experience, and provide suggestions to choose or further revise the design schemes. These tasks cannot be simply dealt with by BIM or other technologies since they require real-life project experience and sufficient intelligence to interpret the estimate results. Therefore, QS will not be marginalised by BIM or AI.

4.2.2 Design-stage cost plan in the context of BIM

4.2.2.1 Identifying the BIM objectives

According to the HKIS (2016), the cost estimate at design stage is called 'cost plan'. An accurate and reliable cost plan is of vital importance to form the basis for subsequent project activities, such as cost planning, tendering system design, construction contracting, and risk management (Southwell, 1970; Skitmore and Marston, 1999; Cartlidge, 2009; Towey, 2012). As more design information surfaces, the methods of cost planning generally range from QTO to tallying building elements from design drawings and assigning cost information to each element to forecast the overall construction cost. This process involves carefully reviewing design drawings from various disciplines at different scales, i.e., the detailed, typical, plan, and top-view drawings to extract the quantities of building components whilst considering the SMM required by local regulations. The QS, however, may occasionally make mistakes when extracting quantities from various design drawings and, consequently, spend considerable time and efforts in checking and adjusting the quantity information. This has triggered the need to bring BIM into the existing QS practice to make cost planning a more accurate and efficient process.

The first objective of BIM at design stage is to develop, enrich, and modify the design BIM model. A design BIM model not only stores and visualises the information depicting designer's intentions but also provides an integrated platform where designers from different disciplines could corporate. The design BIM model should also include all the necessary information for design-stage cost planning. In comparison with preliminary cost estimate, it is much more complicated to develop the detailed cost plan even in the context of using BIM. Some cost-related items, which are key ingredients for estimation, are not explicitly presented in a BIM model, such as finishes, the earthwork, and the formwork and falsework. To overcome this problem, Chen et al. (2015) suggest that the information of non-physical items should also be affiliated with BIM objects by mining historical cost data. Estimators employing BIM will need to develop methods and standards for object development that supports a sufficient LoD required for estimation, as well as provide a framework for guiding consistent information for the BIM components tallied.

Having the BIM model at hand, the next BIM objective is to conduct BIM-based cost estimation by extracting the quantities and assigning them with the cost information. For QS, the challenges of this process include both 'hard' and 'soft' issues. Typical examples of the hard issues include manually adjusting the estimated cost of some BoQ items, designing different methods to quantify different cost items (e.g., directly take-off, derived take-off, manual take-off), and so on. The hard issues require technical and professional QS knowledge to find out a solution. Comparatively, the 'soft' issues concern the managerial aspects, such as designing appropriate procedures, allocating personnel for BIM tasks, and training and education.

4.2.2.2 Mapping out BIM procedures and information flow

At the design stage, BIM-based cost planning will start with a design BIM model (see Figure 4.4). As more detailed information becomes available, the BIM objects of the design BIM will be enriched to LoD 300–350. These BIM objects are not only geographically represented by size, location, orientation, and configuration information but also enriched with non-graphic information such as material, cost, and manufacturer. The design BIM model is generally prepared by architects, structural engineers, and MEP engineers as they are refining their design schemes. Then, QS will further enrich the design BIM with more detailed, particularly cost-related information, to develop a QS-BIM model. The information enriched by QS should include:

- Finishes/building service details,
- Updated prices of cost items,
- Deduction rules based on supplier's pricing, and
- Local QS specifications.

QS should interact with different parties to further enrich the graphic representations of building elements (Aibinu and Venkatesh, 2013). When the BIM model is handed over by designers, QS should start with identifying and specifying any vague design, for example, the ambiguous information on the dimensions, material, finishes, configurations, and connections between different types of building elements. The information should be marked and clarified through query sheets or face-to-face meeting with designers and subsequently updated to the model.

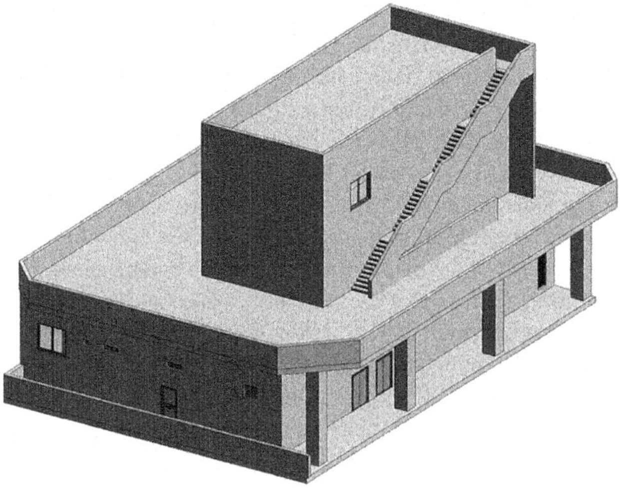

Figure 4.4 A design BIM model.

Then, the updated design BIM model will be linked to estimating software. This can be achieved by either exporting the quantity information to external software, bridging the QS-BIM with the external estimating software, or using the BIM-based estimating software with built-in estimating functions. This process has already been discussed in previous research. For example, Monteiro and Martins (2013) proposed an estimation method that used the dimension and material information in BIM to estimate temporary works or non-explicitly represented items, including earthwork, coatings, and formwork and falsework. Similarly, Kim and Teizer (2014) applied rules to determine the quantity of scaffolding based on the geometry information of structural components in BIM.

QS will also constantly collaborate with the design teams, facilitating them to enrich the BIM model with more detailed and unambiguous information whilst keeping the overall construction cost under the budget ceiling. Based on the model, the BIM tools could help extract quantities, estimate construction cost for the design, and prepare the pre-tender estimate reports, cost plan, and the description and exclusion of price basis.

4.2.2.3 Supporting technologies and QS' responsibilities

Based on the aforementioned discussion, the BIM software should be equipped with the following three functionalities before it could provide the best service to support design-stage cost plan. It is necessary, above all, to have a user-friendly module to facilitate model editing. BIM models from designers, in most cases, cannot be directly applied for QTO or cost estimating since they lack sufficient details for cost-related works. In this regard, the model editing module could help QS to mark or comment on the ambiguous designs, revising and enriching the design and cost information. Besides, the BIM software should be able to extract quantities based on the local SMM and present the quantities in line with the users' needs. In addition, to facilitate an accurate and efficient cost estimation, the QS firm should develop their BIM-based cost database, which can also be linked to the BIM-based software to conduct cost estimate.

Even with the assistance of BIM to relieve QS from tedious QTO works, QS still play an important role in the process. Whilst features of a building design affect the development of detailed estimates, the interpretation of design features relies on the human character of the estimator that could be very different from one to another (Staub-French et al., 2003a, 2003b). To facilitate this process, the schedule of each trade should be integrated with BIM by linking the work items with relevant objects in BIM (Lu et al., 2016). A clear presentation of objects' features with construction sequence helps to estimate the production rate of labour and plant for all work items (Lu and Olofsson, 2014). Besides, QS should also explore a feasible working process involving BIM, assigning suitable methods to quantify different cost items.

4.2.3 *Tendering*

4.2.3.1 Identifying the BIM objectives

At the tendering stage, QS are responsible to prepare tender estimate, tender documents, and tender appraisal. Through rounds of design changes and addendum, the building design should be finalised at this stage. The problems with the existing tendering methods lie in the low efficiency to conduct pre-tender cost estimate, produce the tender documents accordingly based on the locally specified BoQ as well as client's requirements. In addition, a more efficient way to quickly compare BoQ handed in by tenderers is much desired.

BIM is involved in the loop to tailor the aforementioned cost management requirements and make tender process more efficient. Specifically, the BIM objectives at this stage involve finalising the as-design BIM model, conducting BIM-based pre-tendered cost estimate, and preparing tender documents and tender appraisals based on local specifications and client's requirements. Specifically, the as-design BIM model should integrate and update the design information from various stakeholders and provide accurate and timely cost information from the market. The tender documents, such as the items of BoQ, should also be linked to the BIM objects for easy checking.

Ideally, instead of sticking to the traditional DBB process, an IPD model should be adopted with BIM integrated into the QS practice. Here, the BIM-based tender process is mainly based on the existing QS practices and the prevailing procurement methods, with the aim to provide reference for QS companies to explore BIM methods for cost management activities. The contents related to innovative procurement methods such as IPD are presented in Section 3.2.4.

4.2.3.2 Mapping out BIM procedures and information flow

As suggested by the QS-BIM execution plan (see Figure 4.2), the BIM procedure at the tendering stage will start with finalising the as-design BIM model. After negotiations and meetings between different stakeholders, the building design should be finalised with detailed and coordinated information (e.g., architectural, structural, MEP, and building service components) updated to the BIM model. At the time when the model is handed over to QS, QS should further check if there is any missing or vague information necessary to price the BoQ items, and query the designers to fix the problems before construction starts. In a sense, the as-design model should contain the consolidated design information which should be detailed enough to support the subsequent construction process. The BIM objects of the as-design model usually reach LoD 300–400, which are not only graphically represented with detailed representation information such as size, shape, location, quantity, orientation, assembly, and installation but also updated with non-graphic information such as the most updated unit price information based on the market rates.

The as-design model is useful in preparing the pre-tender documents. For example, the pre-tender estimate could be conducted based on the as-design model through a similar process of design-stage cost plan (see Section 4.2.2). Suggested by the HKIS (2012c), it is recommended that tender documents should contain the following documents:

- Instructions to Tenderers;
- Conditions of Tender;
- Form of Tender;
- Conditions of Contract and Special Conditions of Contract (selection of Form of Contract depends on procurement option adopted);
- Preliminaries Specification;
- Technical Specification, selected depends on procurement methods adopted, such as BoQ, Schedule of Quantities and Rates, or Performance Specified Works/Breakdown of the Contractor's rates and prices (for Design & Build contract);
- Schedule of Drawings; and
- Drawings (soft or hard copy).

Some tender documents are prepared by designers, whilst some rely on QS' expertise and experience, for example, Instructions to Tenders, and Conditions of Tenders. BIM tools could facilitate QS to efficiently produce some tender documents. For example, the BoQ and schedules of rates could be prepared by BIM, provided that the requirements of BoQ or schedules of rates are predefined as template embedded in the BIM tools, or specified by QS whilst they are creating the BoQ via BIM. QS shall check all the documents before packing and sending them to potential tenders.

Tender documents generated at this stage could serve as the important reference for tender appraisal. BIM also enhance the efficiency in appraising tender documents from bidders, depending on the format of these tender documents.

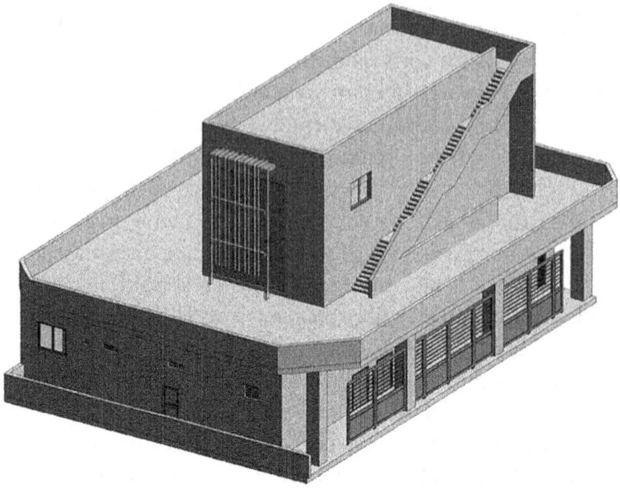

Figure 4.5 An as-design BIM model.

Currently, bidders are required to submit 2D documents, for example, the BoQ and schedule of rates, using the template provided by clients. QS will then compare the tender documents both from bidders and prepared by themselves to generate the tender appraisal report. This is done based on 2D tables. BIM may do a little favour when this form of tender documents is in use. However, in the future, bidders may be required to submit a priced BIM as part of the tender documents (see Figure 4.5). In this case, QS could directly compare the BIM models provided by bidders with their own QS-BIM models to generate the tender documents, tender addendum, and tender appraisal report.

4.2.3.3 Supporting technologies and QS' responsibilities

To provide the best support for the tendering process, BIM software should be embedded with local specifications. Local specifications not only specify the SMM to calculate the quantities but also recommend or regulate the format and contents of tender documents to offer a consistent format applied in the local AEC industry. If the BIM software simply generates the cost and tender documents without considering the local specifications, QS need to spend considerable time and effort to further revise those documents. Therefore, the BIM software should be tailored to those specifications, or offer customisation modules that QS could develop their own templates to cater their needs. QS firms are also encouraged to develop their in-house programs to link with the BIM software. In this way, the firms can generate tender documents for their own needs based on the information from BIM and gradually develop their in-house BIM project database.

4.2.4 Cost control

4.2.4.1 Identifying the BIM objectives

Cost control is one of the core services provided by QS at the construction phase. The primary objective of cost control is to achieve the balance of expenditure between various components/elements of the project and keep final expenditure within the client's approved budget (see also Section 1.2.4). Traditionally, QS will rely on the construction progress reports issued by the contract administrators. The progress reports are summaries of minutes of construction progress meetings and reports received during the meetings. In this regard, it is not an easy job for QS to extract the variations from these reports. Meanwhile, since the reports are often provided monthly, which makes it rather difficult to timely trace the variations and quickly take corrective actions to control the overall construction cost.

BIM should help integrate timely and accurate feedbacks from construction sites and facilitate QS to make corrective actions to control the overall costs. During the construction stage, the contractors should develop an as-built BIM by updating it with actual construction information. The as-built BIM can help track construction process and variations, based on which monthly cost reports can be automated and the interim payment can be settled with better visualisation and traceability.

4.2.4.2 Mapping out BIM procedures and information flow

One of the key challenges of cost control is to get timely and accurate feedbacks from construction sites, managing various types of information, and analysing the information to prepare the cost report. An as-built BIM can help prepare timely and accurate reports by linking to as-built design, schedule, and actual cost. According to Chen et al. (2015), and as-built BIM is an information model with information synchronised with ongoing building process in a real-time manner to reliably and usefully support information exchange and decision-making (Figure 4.6).

Notably, BIM can be also linked to advanced technologies (e.g., laser scanners, sensors, camera), which helps collect on-site data and automatically/semi-automatically update the as-built BIM in a real-time manner. Meanwhile, the BIM-based software will compare the as-built BIM model and the as-design BIM model to identify the variations. Once the variations cross the preset cost thresholds, the software will inform project manager to take appropriate corrective actions that would minimize the cost overrun in a timely manner. The software may also help prepare cost reports (e.g., financial statement, interim payment, and variation reports) if the report format and template are embedded. At this stage, QS regularly prepare cost reports taking into account of anticipated variations, possible claims for loss and/or expense, and fluctuations in the cost of labour and/or materials, with an aim to keep the final expenditure within the clients' approved budget (HKIS, 2012). This process happens repeatedly until the construction project ends.

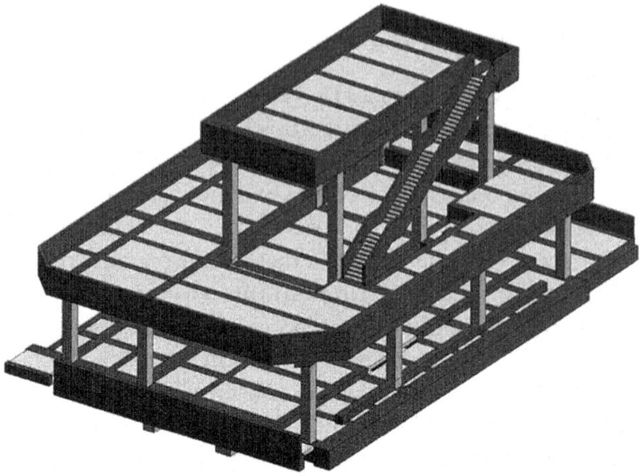

Figure 4.6 An as-built BIM model.

4.2.4.3 Supporting technologies and QS' responsibilities

To acquire the concrete information on the project progress from the sites constitutes one of the major challenges of QS' tasks. Previous studies have introduced many advanced technologies for collecting real-time or near real-time on-site data and integrating it into the virtual BIM model to keep the model updated with the physical construction progress. Chen et al. (2015) pointed out three most popular technologies, namely, laser scanning, radiofrequency identification (RFID), and camera. Specifically,

- Laser scanning technology can capture the geometric data and spatial relationships through laser light. It can be used for process tracking, generation of as-built models, on-site situation monitoring (Shih and Huang, 2006; Turkan et al., 2012), and offering accurate real-life data for QS to make informed decisions.
- RFID, using electromagnetic fields to identify and track tags attached to objects, is suitable for projects using prefabricated components. It can also help to update material and component inventory information in BIM, and trace the real-time consumption of materials and components, thus estimating the real-time expenditure (Lu et al., 2011).
- Video cameras can facilitate tracking the actual construction progress by developing an as-built model based on photos/videos taken from different positions (Chen et al., 2015). Video cameras are now standard equipment installed on-site to detect and track static and moving objects such as workers and equipment on the construction sites. They can help QS better understand the progress on the ground.

With the most updated on-site information, QS should be able to interpret the information, quantify variations, and constantly communicate with the contractor to take corrective actions to control the real-time budget. QS will also participate in regular cost review meetings at this stage. During the meetings, they will try to reach an agreement on variations with the contractor. This is done by exchanging variation reports generated by BIM before the meetings, marking the potential problems and disagreement on some cost items, and discussing the cost items during the meetings to reach an agreement. BIM, with the visual representation of on-site information, could help the parties to better understand the real-time situation of the construction site.

4.2.5 *Variations and final accounts*

4.2.5.1 Identifying the BIM objectives

After construction work completes, QS will settle the final payment and final accounts in a manner similar to the interim payment. This process is done based on the variations between final construction works and the baseline elemental cost plan. An as-is BIM model will be developed to reflect the current as-is condition of the construction project. Then, the variations between the as-is and as-design BIM models will be calculated as the basis for settling final payment and accounts accordingly.

4.2.5.2 Mapping out BIM procedures and information flow

Once the construction project ends, QS should remeasure all the quantities and update the BIM model based on the final drawings, construction reports, and reports of site measurement. Specifically, the variations could include alterations to the design, quantities, quality, working conditions, or sequence of work. These items could change after the settlement of design to cater technological advancements, statutory changes or enforcement, geological anomalies, unavailability of the chosen materials, and change in market conditions. Specifically, the information input includes:

- Final information on building components, for example, their quantities, sizes, locations, materials, and finishes;
- Actual construction methods and schedules; and
- Actual cost of construction like the price of material, labour, and machine.

Then, based on this information, QS will prepare the final report with the help of BIM software to compare the variations between the as-design and the final as-is BIM model (Figure 4.7). Embedded with the final report template, the BIM software may also help generate final cost reports, including variations assessment, payment, and final account. Suggested by the HKIS (2012), the final account will include:

- Architects' instructions;
- Conformation of verbal instructions;
- Prime cost rates adjustment;

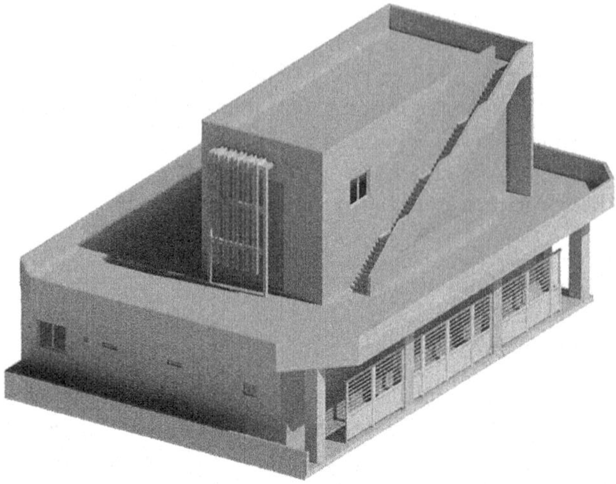

Figure 4.7 An as-is BIM model.

- Prime cost sums adjustment;
- Provisional sums adjustment;
- Provisional quantities adjustment;
- Fluctuations adjustment;
- Contractual claims; and
- Others.

4.2.5.3 Supporting technologies and QS' responsibilities

As construction work is about to finish, it is suggested that QS should summarise issued variations as early as possible to settle the final payment. With the assistance of BIM and advanced technologies tracking the on-site construction progress, QS could quantify the variations much easier by comparing the as-design and the as-is BIMs to measure the changes, omissions, and additions from the models. If the changes are substantial, QS should remeasure the corresponding construction works on-site to finalise the variation quantities. Based on the quantity of variations, QS will further assess the rates of corresponding cost items listed in BoQ to generate the final account. It is suggested that QS should arrange a cost review meeting with contractors to confirm the final account before settling it. The as-is BIM could provide vivid demonstration of the constructed facilities to enable QS, contractors, and other parties to easily understand the constructed facilities and on-site problems, thus consequently enhancing the efficiency of the cost review meetings.

4.3 Critical success factors of BIM adoption for QS

4.3.1 *Demanding a QS-BIM*

In order to adopt BIM for cost management, the first step is to construct a building information model containing the essential details for cost-related tasks. Ideally, the design team (i.e., architects and engineers) should provide the models in a way for QS to extract all the necessary information. In existing BIM-enabled cost management practices, BIM solutions (e.g., RIB iTWO or Autodesk) will normally develop a qualifier to check the models before the extraction is performed. For example, according to RIB, such qualifier is to check BIM model data to ensure quality, allowing the combination of several models (e.g., architectural, structural, and MEP). When there is an error, clash, or non-compliance detected, normally, QS will request clarification from the designer. They believe that QS are serving designers, rather than in a position to replace the designers to design something. In reality, it is not uncommon that QS are only provided with traditional 2D drawings whilst client is eager to harvest the benefits of BIM at later stage (e.g., BIM-enabled estimate, cost control, and final account). Very often, the designers' models do not contain all the necessary information for cost management, especially when the various professionals involved only add contents within their field of interest, for example, architects focusing on the aesthetic and functional aspects, structural engineers on structural safety and durability, and MEP engineers on MEP systems. In any case, it presents an opportunity for QS to demand a QS-BIM, which is a concept on the rise in the QS community around the world but yet to be articulated.

The building information model for quantity surveying (QS-BIM) is intrinsically an information hub that contains the necessary information for quantity surveying tasks. A QS-BIM, above all, is a part of the integrated information hub for a construction project; its information should be included in the integrated BIM. Developing such a model requires reiterative interactions between QS and designers to keep it compliant with existing building codes, updated, buildable, and within the proved construction budget. Besides, the QS-BIM should have a specific lens on quantity surveying domain. The idea embedded is that instead of focusing on all the information presented within a BIM, quantity surveyors should demand the information in BIM that is amenable to QS tasks.

A QS-BIM should contain the explicit, detailed information of the high-value cost items. This information is first developed by architects and engineers when they develop the building designs and may be further enriched with more details. Generally, the project planning parameters in BIM could satisfy the preliminary cost estimate needs. To conduct a detailed cost estimate, comparatively, requires more detailed information, including the architectural, structural, and building services/finishes information. Typical examples of architectural items include doors, windows, and internal walls, whilst the structural items include columns, beams, and slabs. The names of these items should follow a predefined manner such that the items could be automatically extracted, organised, and analysed to generate cost analysis results. Normally, the pre-agreed naming conventions can be found in various BIM standards (Chen et al., 2017).

The QS-BIM should also comply with standard rules of measurement, specifications, and standards used in a specific local area. For example, the Standard Method of Measurement 4 (SMM4), adopted in Hong Kong, stipulates the methods for measuring construction cost by trade. Therefore, information stored within each BIM object should include, along with the standard size and object location data, its material specificities (e.g., types, concrete strength, and configurations) and the connection between adjacent BIM objects for necessary quantity deduction. Here, we particularly highlight the finishes information from the architectural ones, since it will be largely ignored during the design phases, whilst they normally constitute a large portion in the total construction cost. It is unrealistic to demand a high LoD (e.g., 350 or 400) BIM at the very beginning. As previously illustrated, different LoDs BIMs can be used for different estimation purposes (e.g., preliminary cost estimate or design-stage cost plan). A QS-BIM should allow the LoD to increase as the BIM develops in line with the project progression. Figures 4.8–4.10 present the requirements of QS-BIM based on SMM4 in the context of Hong Kong.

More comprehensive requirements of the QS-BIM are subject to further development. Based on the general descriptions of BIM LoD, a specific set of LoDs for the QS-BIM should be developed. We have seen this similar development in other practices. For example, in integrating BIM and GIS, there are endeavours

Item	Operational Step	1. Quantity Surveying Issue	2. Technical Issues	3. Management Issue	Sample Model
Modelling-Architectural	CAD identification	The model is used for 1. (VIII) BRICKWORK AND BLOCKWORK, and (VII) CONCRETE WORKS (concrete non-structural walls), 2. (XIII, XV) WOOD/ STEEL AND METAL WORKS (Door and Window) 3. (XIV) IRONMONGERY can make use of the qty in Golden for its qty 4. (XXII) LANDSCAPING	The first step is to use CAD identification, if the element cannot be identified after checking from the drawings, draw them manually Naming of Door and Window are required	Modeller should write the bill description for (VIII) Brick and Blockwork, (XIII, XV) Wood Works and Steel and Metal Works Modeller should get the quantity for wood works and steel and metal works which require manual takeoff from 7. Manual Taker and those related to room finishes from 5. Finishes Modeller	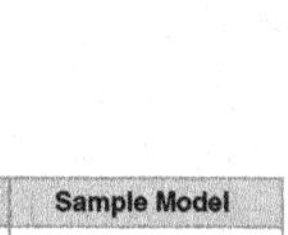

Figure 4.8 Architectural issues in a QS-BIM model.

Item	Operational Step	1. Quantity Surveying Issue	2. Technical Issues	3. Management Issue	Sample Model
Modelling-Structural	a) CAD identification b) Axis c i) Element- Column c ii) Element- Structural Wall c iii) Element- Beam c iv) Element- Slab d) Custom Point and Area e) Precast Concrete Works f) Reinforcement	The model is used for the measurement of (VII) CONCRETE WORKS (except concrete non-structural walls) based on structural drawings provided by structural engineers	The first step is to use CAD identification, if the element cannot be identified after checking from the draawings, draw them manually Naming of Column, Structural Wall, Beam, Slab and Precast Facade are required	Modeller should write the bill description for (VII) Concrete works Modeller should get the quantity for concrete non-structural walls from Architectural Modeller	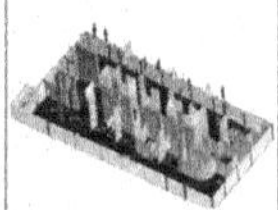

Figure 4.9 Structural issues in a QS-BIM model.

Item	Operational Step	1. Quantity Surveying Issue	2. Technical Issues	3. Management Issue	Sample Model
Modelling-Finishes	CAD identification	The model is used for 1.(XVI)PLASTERING AND PAVING (INTERNAL, External for reference) 2. (XIII) WOOD WORKS(except) those on furniture and fittings 3. (XI) STONE WORKS (ditto) 4. (XV) STEEL AND METAL WORKS (ditto) 5. (XX) GLAZING (ditto) 6. (XXI) PAINTING (ditto) 7. (X) WATERPROOFING (ditto) 8. (XII) ROOF TILING (ditto)	The first step is to use CAD identification, if the element cannot be identified after checking from the drawings, draw them manually Naming of Finishes in item (e) are required	Modeller should write the bill description for (XVI) PLASTERING AND PAVING, (XXI) PAINTING, (XX) GLAZING, (X) WATERPROOFING, (XI) STONE WORKS (XI)	

Figure 4.10 Finishes issues in a QS-BIM model.

to align the two LoDs, which have been separately defined in each domain (e.g., BIMForum, 2017). However, it is not the intention of this book to emphasise the importance of defining and demanding the QS-BIM so that all professions can work together towards a clearer target. QS are facing the challenge to equip themselves with competencies of developing the QS-BIM. It is also an opportunity for the QS profession to elevate their status in the whole project delivery professional ecosystem – a target that has long been due.

4.3.2 Information availability

Another critical success factor of BIM adoption for QS practice is the availability of quantity and cost information. As argued in Sections 2.4 and 2.5, the keyword of BIM is information. The cost information for estimation should involve labour, equipment, and material overhead. Figure 4.11 illustrates an expandable scheme for storing cost information in the database. For labour cost, the hourly rate of each labour trade lives in the database. Each labour trade is assigned with a code for easy identification and searching. For equipment cost, the purchase or rental fee of each type of equipment is stored in the database together with the equipment code and its functions. The rental fee can be updated on daily, weekly, monthly, or even annual basis. For material cost, the measurement unit and unit cost of each material are stored in the database. Similar to the code of labour and equipment, material code should also be recorded in the database.

Once the BIM model is developed, the BIM objects will be linked to the cost database and then calculate the cost of individual BIM objects (Wang et al., 2016). This can be done by linking the BIM object ID and the codes of various cost items. There are generally three types of bidirectional mapping between BIM objects and the cost data (Nassar, 2011). The first one is the one-to-one mapping, where an object in the BIM model can be directly mapped to the cost database. This type

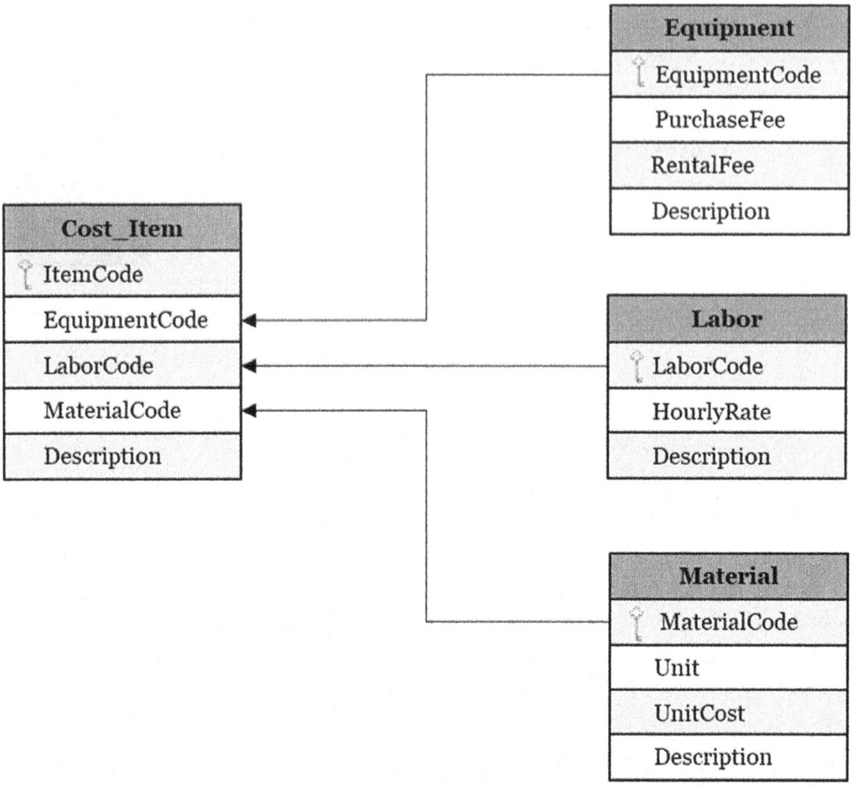

Figure 4.11 A sample of a cost information scheme.

of mapping is typical for fixtures and electromechanical devices. The second type concerns the many-to-one mapping, where many objects have to be mapped to one cost item. For example, the cost of construction equipment like pumps should be divided into a number of relevant objects in the BIM model. Third, the one-to-many mapping entails a BIM object being mapped to multiple items in the cost database. For example, a column has to be mapped to cost items like concrete, reinforcement, and formwork. The quantity surveyor should ensure correct linkages between BIM objects and cost data in order to improve the accuracy of estimation.

Logically, the costs of all working items necessary to construct that component equal to component's total cost. It is like clicking a button and the cost estimate, including the QTO and the BoQ, can be generated automatically. However, in reality, there are many practical issues that should be addressed before the BIM-enabled QS can be realised. For example, BIM should be developed in a way to suit for the QS, for example, different LoDs for different purposes at design stage or tendering stage (referring to Figure 1.1). There are calls for a QS-BIM, and the authors of the book, in later chapter, will join in this call by elaborating how the

QS-BIM should look like. For some finishing, there might be no detailed materials or dimensions specified in a BIM or BIM library. In real-life QS practices, this part should be in accurate terms. Similarly, the materials and man-day of some joints should be deducted rather than simply adding them up together. BIM and the computer programs will not help automatically unless they are clearly instructed by QS.

4.3.3 *Compatible with current QS practices*

Another critical successful factor is that the adoption of BIM, as a facilitating technology, should not significantly disrupt the existing working norms and practices. Rather, it should be carefully designed to be integrated in the norms and practices. The QS-BIM must rope in numerous other issues, such as the BIM technology and process to achieve the cost management goals. A typical example, which has also been mentioned several times in the book, is the SMM4 used in Hong Kong. Over the years, the SMM4 has been developed into a rigid system, and QS invariably have commented that the SMM4 is excessively complicated for both conventional and BIM-based QS practices. Nevertheless, it is not the stance of this book to argue that we should blame the SMM4 and wait for its changes. There is no quick fix of the SMM4 at this point of time. Rather, it is sensible for the two parties, i.e., the QS-BIM and the SMM4, to work towards each other.

Cost management processes exist now as they have for some time, and so QS have become accustomed to a kind of business as usual. The advent of BIM, however, may significantly change standard working practices. This makes it rather difficult for QS to accept, learn, and apply new technologies. Therefore, in designing BIM tools for cost management, it is necessary to take the existing cost management process into consideration. Developers should strive to align the technical tools with prevailing cost management practice, teach QS the connotations, benefits and operating skills of BIM, as well as further adapt BIM to meet QS tasks and needs.

4.3.4 *Compatible with existing BIM-based QS solutions*

Having the QS-BIM at hand, the next step is to extract the quantity information and multiply it by the market price specified by material suppliers or from the previous project database to generate the cost reports. This process is generally completed through BIM-based software or platforms. Autodesk (2007), Eastman et al. (2011), and Wu et al. (2014) classify BIM-based cost management into the following three types.

Exporting quantities to traditional estimating software: According to Eastman et al. (2011), this technique quickly calculates and exports quantity information from the BIM into spreadsheet form as an umping off point for QS' estimation. This may seem lacklustre to most BIM practitioners. However, it is quite simple to apply, control, and align with existing cost management processes such that QS may be more willing to adopt and master it (Autodesk, 2007). However, in the case of addendum, the spreadsheet method may prove problematic as it fails to properly link MS Excel and BIM software. When a design change occurs, inevitable for all projects, quantity surveyors may need to reiteratively and repeatedly go through the process to get a cost report. Therefore, investing in a standardised modelling process that properly connects estimating software, such as MS Excel, with BIM software would lead to superior performance and results (Wu et al., 2014).

Bridging BIM design tools with estimating software: This method does link cost estimating software with BIM design tools, such that cost information automatically updates with each design change. Employing various plug-ins (e.g., Tocoman iLink) connects BIM design tools (e.g., Revit) to facilitate further cost estimation (Autodesk, 2007; Eastman et al., 2011). QS can define the measurement rules via plug-ins. They can also conveniently modify BIM models in the design software, perhaps adding more finishes/building services details based on their professional understanding of building needs. The BIM model provides QS a visual aid to administer greater accuracy (Wu et al., 2014).

Using BIM-based estimating tools: BIM-based estimating tools can facilitate cost estimation and analysis within the BIM model itself. Examples include Autodesk QTO, Glodon, CostX, etc. Similar to the aforementioned approaches, this approach helps QS enrich BIM information for cost management, but without a second platform. It also supports automated quantity extraction, cost report development, visualised cost information, and manual take-off. By embedding costs within the models, BIM-based estimating tools can automatically update cost estimate results when design changes arise.

The successful adoption of BIM for cost management should not deviate too significantly from existing solutions and argue for a totally new solution. It is encouraging to see that BIM vendors are working actively with QS practitioners in fine-tuning their products.

4.4 Summary

Putting BIM into the context of cost management, this chapter developed the related knowledge that prospective executives of QS firms can directly bring into implementation. By following the RIBA Plan of Work as an analytical framework, this chapter explored the prospects of BIM adoption for cost management. BIM seemingly shows promises in every aspect as included in traditional cost management tasks such as preliminary cost estimate, design-stage cost plan, tendering, cost control, and variations and final account. This is followed by the development of a QS-BIM execution plan focusing on the QS practices. The execution plan carefully elaborated the BIM objectives, mapped out BIM procedures and information flow, and discussed the supporting technologies and QS' changing responsibilities. This chapter further explored several critical success factors of BIM adoption for QS. Notably, a QS-BIM is highly desired as a way to articulate the demands of BIM model if it is to be utilised for QS practices. To demand a QS-BIM is also aiming to elevate the status of the QS profession in the whole project delivery professional ecosystem. Successful BIM adoption for cost management is also dependent on other factors, for example, the compatibility of BIM with existing QS practices and software development.

5 Case studies

The aim of this chapter is to elaborate how BIM can be truly implemented in construction cost management by presenting three real-life case studies. As can be seen in Table 5.1, the projects under study vary from composite buildings, house, and teaching facilities. As a whole, these projects involve the adoption of BIM for most QS tasks. Specifically in the selected cases, BIM and related software are adopted to enhance the efficiency of automated QTO, preparing tender documents, and remeasurement of materials in the final account stage. It follows on the QS-BIM execution plan as proposed in Section 3.4 and puts it into the context to discuss the BIM software adopted, the detailed BIM-based QS processes, and the findings and lessons learnt. This chapter is meant to connect to the QS-BIM theoretical framework with real-life cases and provide readers with frontline experiences.

5.1 Case No.1 BIM-based QTO

5.1.1 Project overview

The project is a composite development located on Kwun Chung Street, Hong Kong (see the project information in Table 5.2). The project is developed by Company Alpha, a private real estate developer. The contract sum is HK$96.88

Table 5.1 Brief description of selected case projects

Case No.	*Project type*	*Project phase*	*Applications of BIM*
1	Composite building	Design stage	Generating quantity take-off of concrete, formwork, reinforcement, and finishes
2	Private house with two stories	Tendering stage	Preparing tender documents, such as BoQ, whilst dealing with addendum
3	Teaching facilities and hotel café	Post-contract stage (i.e., construction, variations, and final account)	Remeasuring the quantity of rebar for final account

Table 5.2 Case No.1 project overview

Developer	*Company Alpha*
Size	200 square feet for each flat, 30 stories
Project duration	September 2014 to June 2017
Contract sum	HK$96.88 million

million. The composite building has 30 storeys, including a ground floor (G/F), roof, upper roof, and rooftop. Specifically, the 3rd to 28th floors are standard floors, each equipped with three 200 square feet flats. Such design of the standard floor is very popular in Hong Kong, since the housing price is rising sharply and smaller residential flat units are becoming more popular. The difficulties of design concern fitting all the functional spaces (kitchen, bathroom, and living room) within a single 200 square feet unit. Fewer partitions and wider use of sliding doors are thus observed so that the usage of the space can be maximised.

This case project demonstrates the application of BIM for QTO, especially for the trades of concrete, formwork, reinforcement, and finishes. In this case project, the architect only provides a set of 2D drawings in PDF format rather than BIM models, but QS can quickly build up a QS-BIM model and enrich its information using the built-in modules from BIM-based cost estimate software called Cubicost Takeoff for Architecture and Structure (TAS). Comparing the BIM-based and manual methods for QTO identified, the former as more efficient and accurate, yet there are still some problems QS has to resolve.

5.1.2 *Overview of the adopted BIM software*

Developed by Glodon Company Limited, Cubicost TAS is a 3D-based BIM tool that facilitates QS to quickly develop a BIM model, adding the necessary QS details and automating QTO based on the input model information. Specifically, the software supports direct import of IFC, DWG, and PDF drawings to build up and edit the BIM models efficiently. The QTO by Cubicost TAS is based on the embedded SMM for different countries, such as Mainland China, Hong Kong, the UK, and Singapore, along with the relationships among various BIM objects. The Cubicost TAS also supports viewing calculations and expressions to facilitate rapid result checking, with a potential to reduce errors in practice.

In this project, two teams were responsible for QTO, but employed different methods. One team used the traditional manual method utilising Microsoft Excel, dim sheet, and paper-based calculation, whilst the other team used Cubicost TAS to prepare quantities. The comparison of the results from both manual and BIM-based methods allows a direct understanding of the differences between the two.

5.1.3 BIM-based QTO

5.1.3.1 QS-BIM model development

As mentioned in Section 5.1.1, the architect of this project only provided 2D drawings in PDF format rather than a BIM model. In this regard, QS needed to first sketch a BIM model in Cubicost TAS before taking off the quantities.

The built-in modules adopted Cubicost TAS, first by importing the PDF/CAD files of the building design drawings into the software. After that, Cubicost TAS helped identify layers in the PDF/CAD drawings and generate BIM models accordingly. The principle of building up the model is stretching the layers from the plan drawings to certain indicated elevations; BIM objects from the same layer were grouped under the same type. QS then associate dimensions to the corresponding BIM objects. Notably, the BIM models generated based on the PDF files will have some information or even layers lost, whilst the BIM models generated based on CAD files are more accurate and efficient (i.e., around five times faster than the PDF methods). The scaling of drawings will also influence the accuracy in developing the BIM model.

To further refine the BIM model for QTO, the next step involves editing each BIM object. QS either rename the BIM objects for better identification purposes, or edit and enrich their attributes (e.g., size, materials, unit price, or finishes information) to make the QTO more accurate. If QS cannot find an appropriate object, they need to use drawing tools embedded in Cubicost TAS to draw the object (see Figure 5.1).

5.1.3.2 Quantities extraction

Having the QS model at hand, QS then extracted the quantity information according to the description in Bills. To do so, they needed to first specify the requirements according to their needs and those of Hong Kong's Architectural

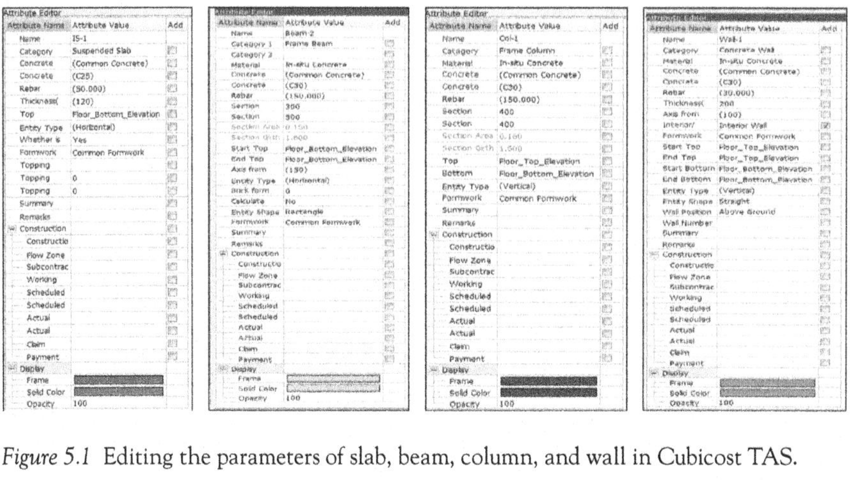

Figure 5.1 Editing the parameters of slab, beam, column, and wall in Cubicost TAS.

Services Department (ArchSD) Model Bills. Being flexible, Cubicost TAS can suit any user or ArchSD Model Bills conditions. The software used a 'view by category' concept as shown in Figure 5.2.

Taking the procedure of measuring concrete work as an example, it generally involves the following seven steps:

- Prioritise the attribute according to Model Bills (by moving Up/Down the attribute names to Prioritise);
- Separate quantities by materials (e.g., glass, rebar, or concrete);
- Within the information of a material (e.g., concrete), separate quantities into Grades (e.g., C30, C60);
- Separate quantities by entity types (i.e., vertical/curved type wall because formwork for straight and curved walls need to be separate according to SMM4);
- Separate quantities by thickness (e.g., 100 mm, 200 mm);
- Separate quantities by floor and component name, although this may not be useful for billing, it is important for tracking quantities and checking (e.g., finding the exact quantity for Wall A); and
- Since Cubicost TAS has many quantity items to be extracted (e.g., volume or area), QS need to select the quantity items according to the existing regulations and practices.

The results of QTO are presented in Figure 5.3.

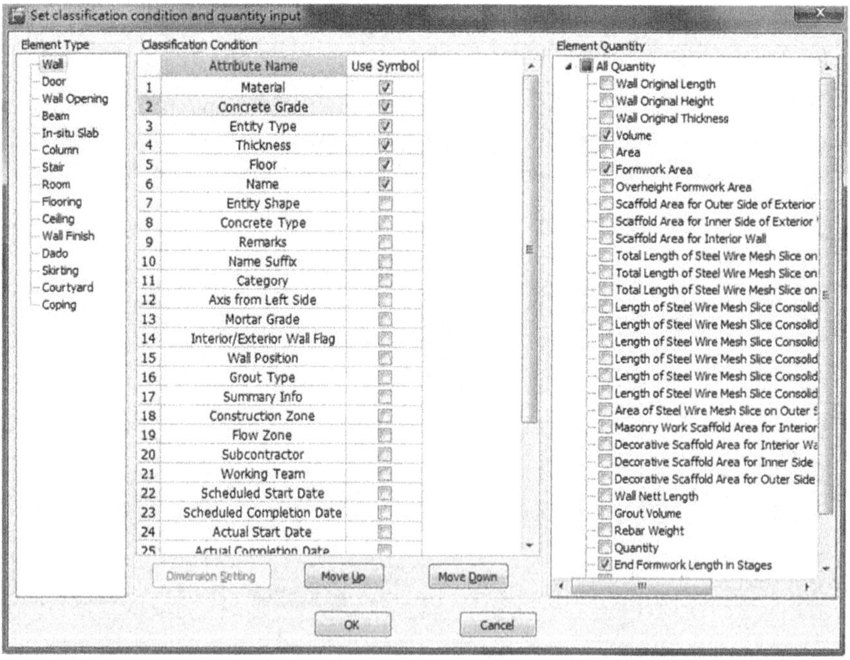

Figure 5.2 Set classification conditions and quantity input for each type of element.

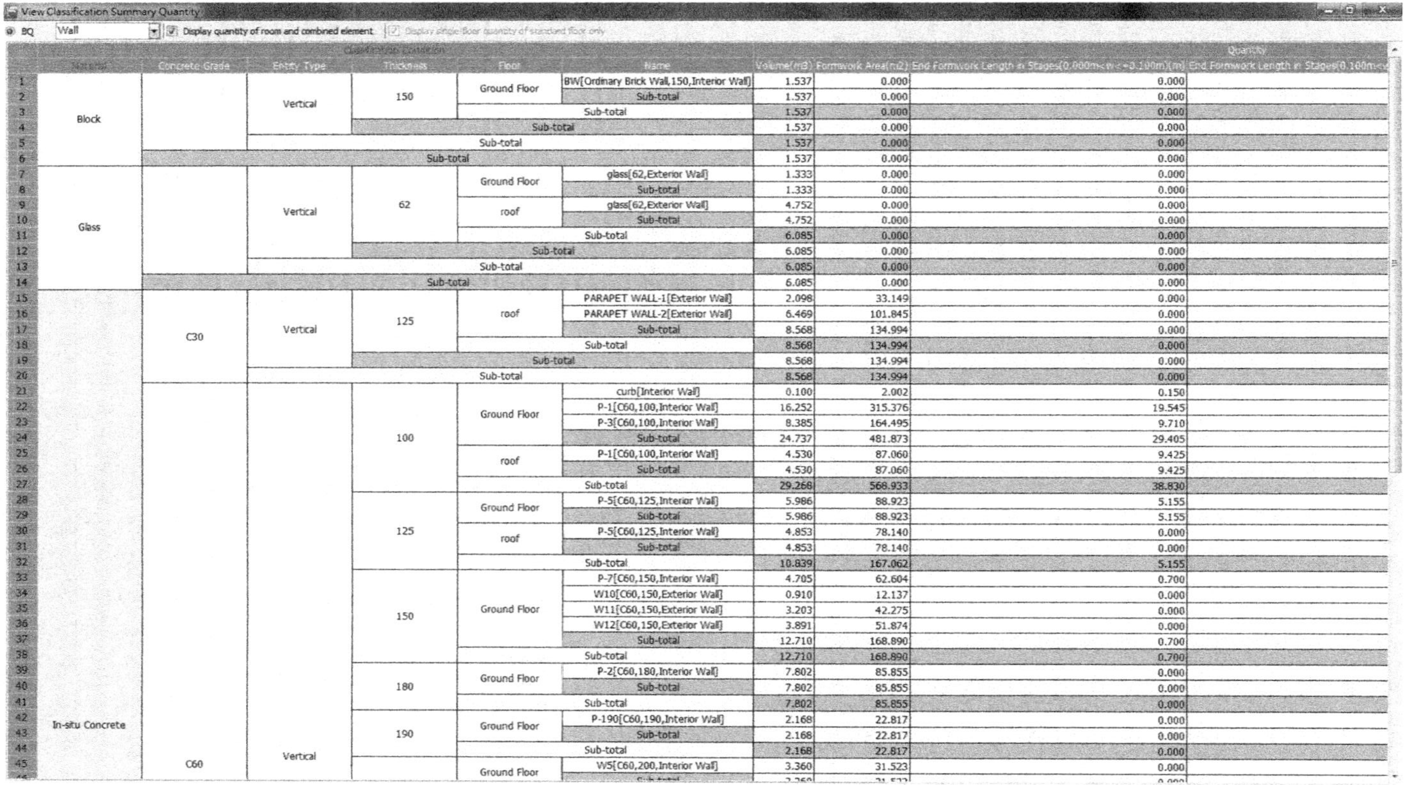

View Classification Summary Quantity

BQ | Wall | Display quantity of room and combined element | Display single-floor quantity of standard floor only

	Classification Condition						Quantity			
	Material	Concrete Grade	Entity Type	Thickness	Floor	Name	Volume(m3)	Formwork Area(m2)	End Formwork Length in Stages(0.000m<w<=0.100m)(m)	End Formwork Length in Stages(0.100m<…
1	Block		Vertical	150	Ground Floor	BW[Ordinary Brick Wall,150,Interior Wall]	1.537	0.000	0.000	
2						Sub-total	1.537	0.000	0.000	
3					Sub-total		1.537	0.000	0.000	
4				Sub-total			1.537	0.000	0.000	
5			Sub-total				1.537	0.000	0.000	
6		Sub-total					1.537	0.000	0.000	
7	Glass		Vertical	62	Ground Floor	glass[62,Exterior Wall]	1.333	0.000	0.000	
8						Sub-total	1.333	0.000	0.000	
9					roof	glass[62,Exterior Wall]	4.752	0.000	0.000	
10						Sub-total	4.752	0.000	0.000	
11					Sub-total		6.085	0.000	0.000	
12				Sub-total			6.085	0.000	0.000	
13			Sub-total				6.085	0.000	0.000	
14		Sub-total					6.085	0.000	0.000	
15	In-situ Concrete	C30	Vertical	125	roof	PARAPET WALL-1[Exterior Wall]	2.098	33.149	0.000	
16						PARAPET WALL-2[Exterior Wall]	6.469	101.845	0.000	
17						Sub-total	8.568	134.994	0.000	
18					Sub-total		8.568	134.994	0.000	
19				Sub-total			8.568	134.994	0.000	
20			Sub-total				8.568	134.994	0.000	
21		C60	Vertical	100	Ground Floor	curb[Interior Wall]	0.100	2.002	0.150	
22						P-1[C60,100,Interior Wall]	16.252	315.376	19.545	
23						P-3[C60,100,Interior Wall]	8.385	164.495	9.710	
24						Sub-total	24.737	481.873	29.405	
25					roof	P-1[C60,100,Interior Wall]	4.530	87.060	9.425	
26						Sub-total	4.530	87.060	9.425	
27					Sub-total		29.268	568.933	38.830	
28				125	Ground Floor	P-5[C60,125,Interior Wall]	5.986	88.923	5.155	
29						Sub-total	5.986	88.923	5.155	
30					roof	P-5[C60,125,Interior Wall]	4.853	78.140	0.000	
31						Sub-total	4.853	78.140	0.000	
32					Sub-total		10.839	167.062	5.155	
33				150	Ground Floor	P-7[C60,150,Interior Wall]	4.705	62.604	0.700	
34						W10[C60,150,Exterior Wall]	0.910	12.137	0.000	
35						W11[C60,150,Exterior Wall]	3.203	42.275	0.000	
36						W12[C60,150,Exterior Wall]	3.891	51.874	0.000	
37						Sub-total	12.710	168.890	0.700	
38					Sub-total		12.710	168.890	0.700	
39				180	Ground Floor	P-2[C60,180,Interior Wall]	7.802	85.855	0.000	
40						Sub-total	7.802	85.855	0.000	
41					Sub-total		7.802	85.855	0.000	
42				190	Ground Floor	P-190[C60,190,Interior Wall]	2.168	22.817	0.000	
43						Sub-total	2.168	22.817	0.000	
44					Sub-total		2.168	22.817	0.000	
45					Ground Floor	W5[C60,200,Interior Wall]	3.360	31.523	0.000	

Figure 5.3 The results of QTO.

5.1.4 Findings and lesson learnt

In this case project, a BIM new-hand used the BIM-based method, whilst two young QS formed a team to manually take off the quantities. A comparison of the results of these two methods is summarised in Figure 5.4.

One of the significant benefits of using BIM is that it achieves a higher level of accuracy for quantities. The BIM method performs much better than the manual method in high value and conventional cost trades like concrete, wall, slab, and beam. The BIM software can compare two different models to calculate quantities. For instance, formwork of structural wall and partition wall made of concrete should be cast together, and the formwork intersecting them should be deducted. Since they are drawn by two different parties, it is very difficult for humans to find the 'exact' dimension overlapping both drawings and deduct the 'exact' formwork area. The BIM software, embedded with local measurement rules, can enable an accurate QTO, which strictly follows measurement rules. The BIM method can also accurately deduct overlaps between adjacent BIM objects based on their relationships, which is hard to be considered when using manual methods and is thus often roughly handled or even ignored in traditional manual methods. Besides, with the information integrated in a visualized 3D platform, BIM makes it easier to check the results of QTO. QS can also manually modify the rules of measurement to ensure the results are satisfactory.

Indicator

	Color showing >10% error
	Glodon bugs
	Glodon remeasurement

Element	Entity	Manual(A)	Rough Manual remeasure(B)	Glodon(C)	% difference (C-A)/A	% difference (C-B)/B	Reason	Comment
1.Column	Concrete	18.89		18.20	-4%		Manual did not minus slab thk	
	Formwork	209	150	146.138	-30%	-3%	Manual did not minus slab thk, Manual height of fwk all set to 5.95m,remeasure no error	
2.Wall	Concrete	924.51	563	520	-44%	-8%	Manual did not minus slab, beam, the timing for 28-R/f shd be 1 instead of 19	Glodon should be able to separate different grade of wall and type of wall, say block wall, parapet wall, etc.
	Formwork	971	4895	4416	355%	-10%	Manual did not times two for wall formwork, the timing for 6-18th is 19 instead of 1	Glodon should be able to separate formwork for different type of wall, say block wall, parapet wall, etc.
3.Slab	Concrete	220		290	32%		Manual did not measure the slab according to semm4, rough remeasurement not possible as too many beams to encounter	Glodon should be able to separate concrete for slab and landing
	Formwork	1612		1617	0%		No error	Glodon should be able to separate formwork for slab and landing
4.Beam	Concrete	280	173	187	-33%	8%	Manual did not deduct slab thk and column	
	Formwork	2454	1549	1436	-41%	-7%	Manual 2nd floor copy 2B16-2B34 twice, did not minus slab	Bugs in Glodon Formwork separation of different height
5.Staircase	Concrete	98		98.5	1%		No error, mass concrete measured separately manually	Glodon should separate mass concrete and RC concrete
	Formwork-landing soffit	126		128	2%		Bugs occur in Glodon as no bottom fwk measured for 6th-28th floor, remeasurement using typical quantity from 3/F got 128	Glodon miss formwork in landing soffit
	Formwork-sloping soffit	310		278	-10%		Manual not deducted beam fwk under stair landing, Manual separate mass concrete	Not possible for human to calculate sloping fwk that accurate, Glodon should separate mass concrete and concrete
	Formwork-riser	900		980	9%		Manual separate mass concrete	Glodon do no show detail on the measurement of riser, Glodon should separate mass concrete and RC concrete
Total	Concrete	1179.89		1119.70		6%	The total between manual and Glodon is close subtracting two factors: 1.remeasurement of wall 2.exclude extra beam fwk	
	Formwork	9601		9001.14		6%	The total between manual and Glodon is close subtracting two factors: 1.remeasurement of wall 2.exclude extra beam fwk	

Figure 5.4 Difference between BIM-based and manual method.

Another significant benefit is that using BIM for QTO can save time and labour force. In the case project, the BIM-based method reduced the time and labour force required to extract the quantity information even considering the extra time spent on developing the model and modifying its information. BIM can help save time and labour force in the following two ways. First, it considers several views of a design simultaneously when extracting quantity information, whilst human beings tend to spend a considerable time extracting the same information by comparing several design views. Second, in BIM software, the design details and cost information can be visualized graphically in a 3D model. This makes it more intuitive for QS to understand the building design and acquire information quicker.

However, the BIM-based method for automated QTO still faces several challenges. The first challenge relates to the development of a QS-BIM model. When developing a model based on 2D drawings using BIM-based cost estimate software, QS need to solve the problems, including the data loss, failing to identify an object due to chaos layer settings in CAD drawings, modifying the BIM object names and classifications, and adding more cost-related information to the BIM model.

The second challenge is that there is still no clear strategy to integrate BIM with the existing QS practices, in particular the procedure to conduct BIM-based QTO within an organisation. The procedure should be detailed (e.g., how to allocate the QS and BIM method human resources, the detailed steps in doing a BIM-based project, or the methods for checking the results generated by BIM). Besides, the QS firms should determine a template to present the QTO, which will be used in the subsequent QS steps.

The third challenge is that, BIM software, like Cubicost TAS used in this case project, was programmed to stick strictly to the rules in a SMM, in this case HKSMM4. This may cause the problem of inflexibility. A good example of this could be how human QS will consider more than just the accurate measurement. There are in fact many other factors that could influence cost during the construction life cycle. The manual method does not 100% follow HKSMM4 in practice, and the quantity measured under manual method tends to be a little bit 'larger' than the actual quantity so as to arrange a buffer for cost control. "Estimating is an art" (Stewart et al., 1995, p. 648), whilst measurement should theoretically be a science. Differences between computer and human judgements are inevitable.

5.2 Case No.2 BIM-based tender document preparation

5.2.1 Project overview

This project involved the construction of a private house with two storeys. It began in October 2014 (see Table 5.3 for more project information). Each floor is around 1,500 square feet. Outside the house, there are a swimming pool and a garden. This project is developed by Company Beta. The contract sum is around HK$85.61 million. The project was started in 2014, and a QS model was built for measurement. At the start of this project, HK$3 million budget was served for

Table 5.3 Case No.2 project overview

Developer	*Company Beta*
Size	1,500 square feet per floor, two storeys
Project duration	October–December 2014 (tender document preparation only)
Contract sum	HK$85.61 million

the application of BIM, for example, clash detection, project coordination and communication. In this project, BIM was adopted for various trades, such as concrete, formwork, and reinforcement. The bulk check method was a traditional bulk check table carried out by a Chartered QS. However, the most difficult part for QS is to create a systematic work procedure for BIM application, for example, how to check the model accuracy and completeness for measurement and how to train up junior staff to do BIM-based measurement.

This case study demonstrates the use of BIM along with traditional cost estimating methods to prepare tender documents from the detailed design stage to the tendering stage. The BIM team for this project consisted of two members (i.e., a senior QS and a junior QS with BIM knowledge, referred to as BIMer in this section). The BIMer was the major executor of the BIM-based methods, in charge of the BIM model development, model information enrichment, especially the finishes information, setting up the schedule formats in BIM software, and extracting the quantities accordingly. The senior QS, comparatively, is more concerned with providing guidance and suggestions to determine the standard schedule formats, revising extracted quantities, bulk checking, and offering QS knowledge to the BIMer. Working collaboratively, the BIM team should also identify potential problems and propose future development strategies for the firm to adopt BIM on a larger scale.

Preparing the tender documents (e.g., the BoQ) can prove more complicated than merely taking off the quantities. BoQ require a more rigid format and contain more information than the QTO, such as the description of each cost item. A standardised BoQ template should be used to reduce the chance of omitting or duplicating the cost items that lead to chaos in tender documents. Addressing the aforementioned problems, Case No.2 not only explores the technical issues (e.g., setting up the standard BoQ template, linking the model with the template, and automatically generating BoQ accordingly) but also the managerial process, BIM procedure, personnel allocation, and interaction between other stakeholders like architects and engineers.

5.2.2 Overview of the adopted BIM software

There are two main tools adopted in this case (i.e., the Cubicost TAS developed by Glodon to extract the quantities and an in-house computer system to further develop tender documents like the BoQ). The Cubicost TAS reflects a BIM-based cost estimate software with a module for QS-BIM model development built-in. Interested readers can return to Section 5.1.2 for a detailed introduction of Cubicost TAS.

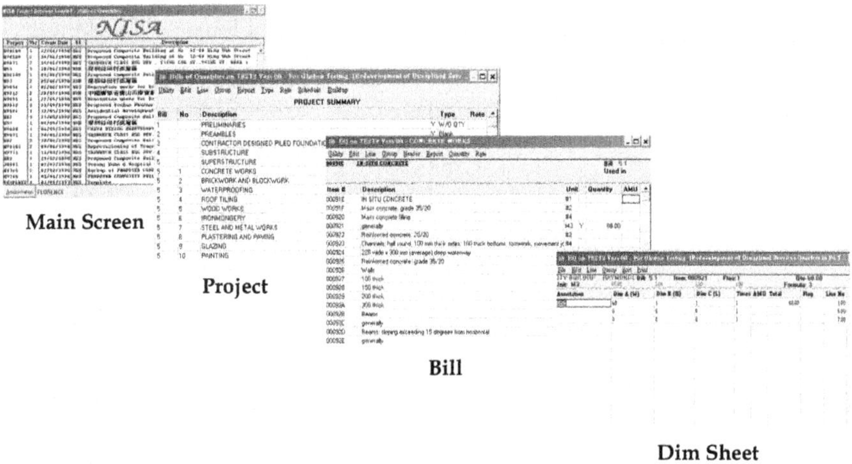

Figure 5.5 System hierarchy and user interface of the in-house computer system.

The other tool, namely, the in-house computer system, developed in the 1990s, is an integrated platform for managing quantity and cost information for each project from pre- to post-contract stages. It allows the collection, organisation, storage of information, and generation of documents (e.g., schedules, BoQ, and contractual documents) based on the templates developed by the QS firm along with the clients' requirements. Specifically, the system consists of ten modules covering the QS firm's core business, such as cost estimate, BoQ, tender preparation, tender appraisal, and post-contract. The system hierarchy and user interface are presented in Figure 5.5.

Designed in a rigid manner to ensure accurate and organised information, the input process demonstrates some clear advantages. However, the bill items developed using this method do not link to the quantities in the QTO documents. Therefore, QS need to manually update the bill items in case of any design change. Given this situation, a more intelligent method is desired to automatically link to design documents, QTO documents, and the computer system to generate tender documents.

5.2.3 *BIM-based tender document preparation*

5.2.3.1 Determining the standard form to generate BoQ

In this project, the QS firmly integrated BIM with their traditional methods (i.e., the in-house developed computer system to enable more efficient tender document preparation). The integration of BIM with traditional cost estimate software starts from QTO via Cubicost TAS. This process is similar to Case 1. The reader may review Section 5.1.3 to refresh their understanding of this process. Notably, the extracted quantities are organised and presented in several schedules based on

Classification Condition					
Concrete Grade	Remarks	Classified	Thickness	Floor	Name
C25			120	**Ground Floor**	**IS-3**
					Sub-total
				2nd Floor	**IS-1**
					Sub-total
				6th Floor	**IS-1**
					Sub-total
				7th Floor	**IS-1**
					Sub-total
				Upper roof	**IS-1**
					IS-2
					Sub-total
				Sub-total	
			150	**Ground Floor**	**IS-1**
					Ramp
					Sub-total
				1st Floor	**IS-1**
					Metal Deck
					Sub-total
				3rd Floor	**IS-1**
					Sub-total
				4th Floor	**IS-1**
					IS-3
					Sub-total
				5th Floor	**IS-1**
					Sub-total
				Roof	**IS-1**
					Sub-total
				Sub-total	
			200	**Ground Floor**	**Lift pit**
					Sub-total
				4th Floor	**IS-2**
					Sub-total
				Sub-total	
			225	**Ground Floor**	**IS-2**
					Sub-total
				Sub-total	
			Sub-total		
		Sub-total			
	Sub-total				
Total					

Figure 5.6 Standard format of the schedules of concrete works.

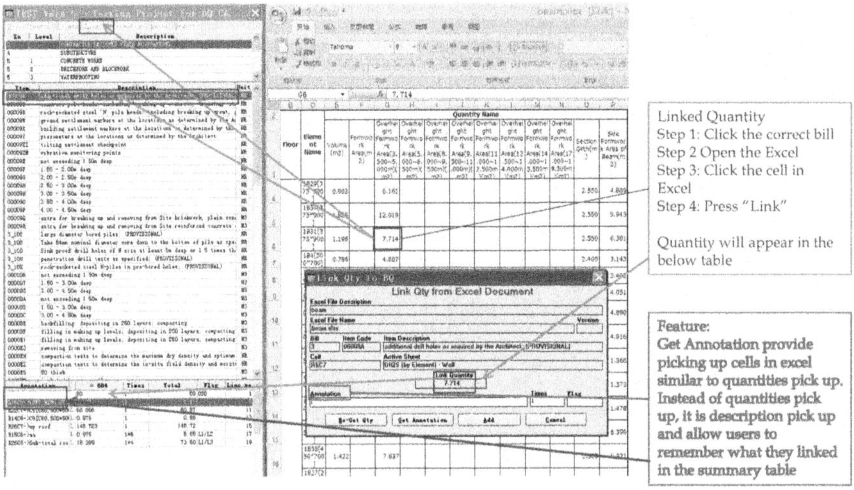

Figure 5.7 Linking cost items of each Bill to quantity information from Excel file.

a predefined standard format to enable convenient linkage between the schedules and in-house computer system. For example, the schedules of concrete works are grouped and presented according to the concrete grade to meet requirements of BoQ format (Figure 5.6). This can be done through settings in Cubicost TAS.

Then the extracted quantities will be imported into the in-house computer system to further generate BoQ. To do so, QS first opened the *Bills of Quantity* module and select the project to activate the Bill editing interface. The interface presents bills that are necessary to generate BoQ in Hong Kong, for example, substructure, concrete works, brickwork, and blockwork. Then QS manually linked the bills with the corresponding Excel files. Notably, once the Excel files are stored, they cannot be directly edited in the computer system. In case of any measurement mistakes or addendum, QS should prepare another Excel schedule with the same file name, so that the newly added file can be stored in the system in different version. QS subsequently linked the cost items of each bill to the cell presenting the quantity information in the corresponding Excel file (see Figure 5.7). Only after all the bill items were filled and linked to Excel cell can the computer system generate the BoQ for tender purpose.

5.2.3.2 *Bulk check*

Boasting the schedules, the next step is to check the accuracy and liability of the quantities extracted by Cubicost TAS. In this project, QS used traditional 'bulk checking' methods. QS rearranged the elements in the bulk check table for a convenient comparison of different indicators (see Figure 5.8). Differing from the schedules where elements are arranged for the convenience of billing, the bulk checking table is usually arranged by floor for the convenience of checking

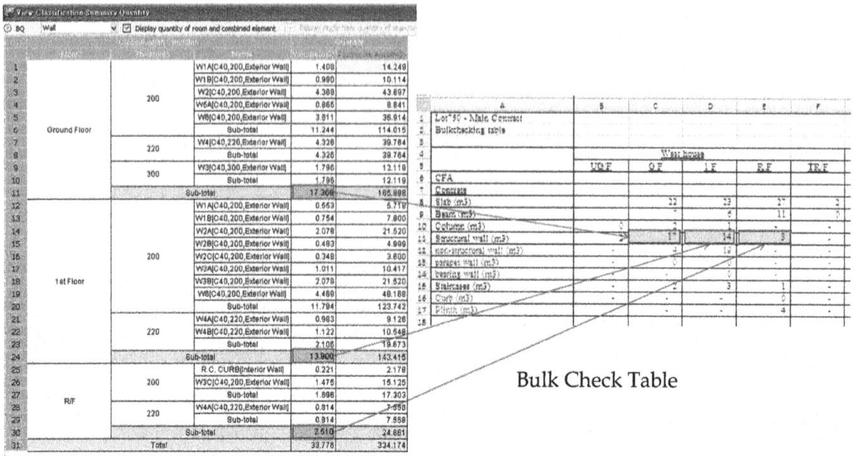

Figure 5.8 Bulk check table of Cubicost TAS.

quantity abnormalities. For example, to investigate the slab quantities, QS could compare the different slab quantities used for each floor. The steel ratio, confirmed by the structural engineers, can also serve as an indicator as it is should mimic the same elements within the given building design.

5.2.3.3 Addendum

During the detailed design stage, QS usually need to interact with other parties to get detailed and updated information for accurate cost estimation. For example, once QS identify any ambiguous or undefined information, they need to query designers for more comprehensive information. The designers also constantly change their drawings due to design mistakes, changing client requirements or desires, or clashes between architectural, structural, and MEP drawings. These situations incur addendums to the cost estimate and thus tender document revision.

There were in total three addendums in Case No.2. The first addendum was a change in drawing clarifying some mistake made in the original drawings (e.g., mismatching of architectural and structural drawings). Similar to other projects, drawings were rarely completely accurate until the very final stage of construction. Mistakes included section mismatching with the plan, rebar details mismatching the framing plan, description of drawings mismatching dimensions on drawings, and so on. When encountering these situations in a 2D environment, query sheets were used. The process should be similar when applying BIM. Occasionally, QS have to assume quantities when doing QTO from BIM models since discrepancies exist amongst parties. These assumptions should be accurately recorded so that QS can extract the right quantity from the model when the queries were clarified with more accurate information.

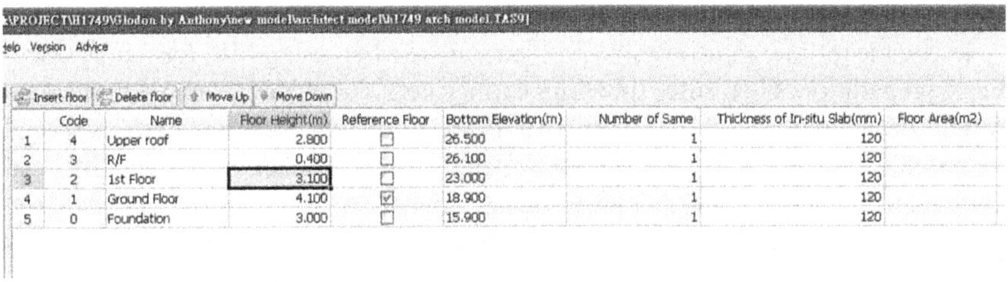

	Code	Name	Floor Height(m)	Reference Floor	Bottom Elevation(m)	Number of Same	Thickness of In-situ Slab(mm)	Floor Area(m2)
1	4	Upper roof	2.800	☐	26.500	1	120	
2	3	R/F	0.400	☐	26.100	1	120	
3	2	1st Floor	3.100	☐	23.000	1	120	
4	1	Ground Floor	4.100	☑	18.900	1	120	
5	0	Foundation	3.000	☐	15.900	1	120	

Figure 5.9 Amendment in model for addendum 2.

Dealing with variations in a 3D model is often different from that in 2D drawings. Addendum 2 as shown in Figure 5.9 demonstrates this idea, where designers change the height of headroom from 3.5 to 3.1 m. When extracting cost from 2D drawings, QS need only to adjust the height information, paying little attention to the position of each building element since it will have little influence over the estimate accuracy. Comparatively, in a 3D environment like BIM, QS need to be especially careful about the coordination (i.e., X-, Y-, and ZX-axes) of building elements because it will affect QTO results in BIM. For example, if QS only adjust the floor height, the vertical position of building components (e.g., beams) will not change with the changing floor height. In this situation, the resulted quantities for formwork would be incorrect.

In addendum 3, a bigger obstacle was encountered. Only scanned hard copies were sent to QS due to time pressures, and some of the features were only briefly described and marked in pen on the hard copies. In this situation, QS had to try their best to build the model without extensively referencing the drawings and accuracy decreased.

5.2.4 *Findings and lesson learnt*

Case No.2 presents some standardised procedures and formatting for BIM-based tender document preparation. In this project, both BIM and the conventional QS methods were adopted for the following reasons. First, the QS firm is still at the early stage of BIM implementation, it is difficult and costly to totally shift their comfortable working procedure, which has been in place for a while. Besides, the current BIM-based cost management software cannot fully support the generation of tender documents based on the local regulations and clients' needs. It will take considerable effort to further revise the outputs if totally relying on the BIM software.

The bulk check table and standard schedule developed in this project were recognised to be effective for tender document preparation. Some benefits of BIM, in this case project, are similar to those presented in the first case (see Section 5.1.4). For example, BIM can improve the accuracy and efficiency of quantity measurement. Furthermore, BIM can reduce the time necessary to prepare tender documents (e.g., BoQ). This is because the quantity of most building components in BIM can be automatically updated as design changes.

In the meantime, some notable problems arose when using BIM for tender document preparation. First, since drawings cannot be 100% accurate, some dimensions should be assumed when developing the BIM model. All such assumptions should be carefully recorded so that the model can be rebuilt after the ambiguities are clarified using queries. Second, when using conventional paper-based methods, QS does not need to worry about the position of building elements. However, things proceed differently when the BIM software is used. For example, when the height of the wall is changed, the upstand beams would stay in the original position in the BIM model, leading to an inaccurate quantity of the formwork. Therefore, QS has to take extra care to check if all the changes are correctly applied in the BIM model. Third, the BIM model alone might not be enough to prepare all documents needed for tendering. QS still need additional software for tendering document preparation, such as the development of an item code. Therefore, the integration and data interoperability of the required software should be carefully considered.

5.3 Case No.3 BIM-based remeasurement

5.3.1 *Project overview*

This project concerns the post-contract stage of a hospitality teaching and training centre managed by a major Hong Kong university and located adjacent to a hotel partially operated by the same centre's students. Gamma Teaching Hotel built with support from Company Alfa is a critical element in the overall plan of the university's business school. Gamma Teaching Hotel allows students to run their own café within the teaching facilities. The facilities feature regular teaching facilities including offices, lecture theatres, computer labs, and a demonstration kitchen. By participating in the daily operation of the demonstration kitchen, students gain more comprehensive understanding of both food production techniques and business theory as they experiment with different business plans and operations.

The post-contract stage of the project started in 2015. The overall project totalled around HK$220 million (Table 5.4). This project demonstrated how BIM could be adopted in the post-contract stage, with a focus on rebar remeasurement. Generally speaking, rebar is one of the most difficult trades to measure and remeasure. Compared to other trades such as concrete and formwork, it requires a comprehensive understanding of various design documents, such as General Notes,

Table 5.4 Case No.3 project overview

Developer	*Gamma Teaching Facilities*
Size	107,639 square feet
Project duration	May–June 2015 (post-contractual stage only)
Contract sum	HK$220 million

Typical Design, as well as specifications (e.g., British Standard) to conduct an accurate rebar measurement. The difficulties of BIM-based rebar remeasurement lie in integrating the necessary design information from various types of documents into a single model in an accurate and efficient manner.

Case No.3 provides a feasible solution to the aforementioned difficulties at post-contract stage. Specifically, this case project presents a working procedure integrating BIM and the traditional QS practice, the selection of BIM-based or manual methods for rebar remeasurement and the QS responsibilities involved. The QS firms may get some insights from this case on how to develop BIM methods and incorporate BIM into existing QS practice, particularly when the AEC industry lies in the early stage of shifting the traditional 2D-based solutions to 3D BIM-based ones.

5.3.2 *Overview of the adopted BIM software*

The BIM software adopted in this project was Cubicost TRB. Cubicost is an integrated BIM solution developed by Glodon for construction cost management during various project phases. Cubicost TRB is one of its most important products and is compatible with other Cubicost products (e.g., Cubicost TAS and Cubicost TME for mechanical and electrical). It attempts to offer a one-stop solution for QTO for numerous trades.

Cubicost TRB was chosen for this project due to its capabilities to provide a quick and relatively accurate BIM solution. Cubicost TRB could automatically identify the rebar details specified in 2D design documents in different formats. Glodon pragmatically recognises that most documents handed to clients and QS are in a 2D format. Besides, a BIM model created in another BIM-based software (e.g., Revit or Cubicost TAS) can also import into Cubicost TRB. This makes it possible for one-time modelling to support QTO for different types of both rebar and other architectural and structural components. In addition, Cubicost TRB offers a user-friendly editing module for users to edit the attributes of rebar.

Cubicost TRB can also perform accurate QTO based on the local methods of measurement. It is embedded with QTO rules and verified by users for several countries/regions (e.g., China, Singapore, Hong Kong, and the UK). Users could also define their own rules of measurement or deduction rules based on the contract requirements and users' needs. It has become easier for users to check the extracted rebar quantities when provided with vivid and integrated 3D presentation of rebar layout and configuration in a BIM. In addition, Cubicost TRB supports BoQ generation in a variety of built-in report formats or user-defined formats.

5.3.3 *BIM-based remeasurement of rebar*

5.3.3.1 BIM model development

In this project, the primary aim of BIM adoption was to enhance the efficiency of rebar remeasurement and the preparation of the post-contract documents. The first step was to develop a BIM model for the BIM-based rebar remeasurement.

Ideally, this BIM model should be prepared at the design stage, updated as construction works progress, and used to analyse the quantities consumed on-site to settle down the final account. In this case, however, no BIM model had been developed before the pre-contract stages since BIM was less established at the time the project began.

Given this situation, QS developed a BIM model based on the issued 2D 'drawings for construction'. These 2D drawings had already been modified to make the design buildable whilst the overall construction cost still could be kept under the approved budget. In this project, only the framing plan was in CAD format, whilst the others, PDF. This made it comparatively more difficult to automatically develop an accurate QS-BIM model, as some layer information was missing. Notably, the BIM model for rebar, especially those complex connections between the superstructure and foundation, was not 100% correct. Besides, the 2D drawings did not necessarily contain all the details for rebar remeasurement. Therefore, it is critical to manually amend the BIM model by QS to foster greater detail or, in some cases, redevelop the digital representation of rebar altogether.

The manual amendment of BIM model relies on the comprehensive understanding of design documents (e.g., General Notes, Typical Design, and British Standards), as well as the software environment concerning the model setting, BIM object attribute edition, and remeasurement rules and rationales. The British Standards relating to rebar measurement regulate the measurement methods of different types of rebar. The General Notes set up the overall details on rebar (e.g., the rebar grade and the concrete cover of different structural elements critical for rebar measurement), whilst the Typical Design further elaborates the detailed layout of rebar within the structural elements. BIM could help QS gain an overall understanding of the design documents, automatically identifying the BIM objects based on 2D drawings, providing 3D views of models for a vivid

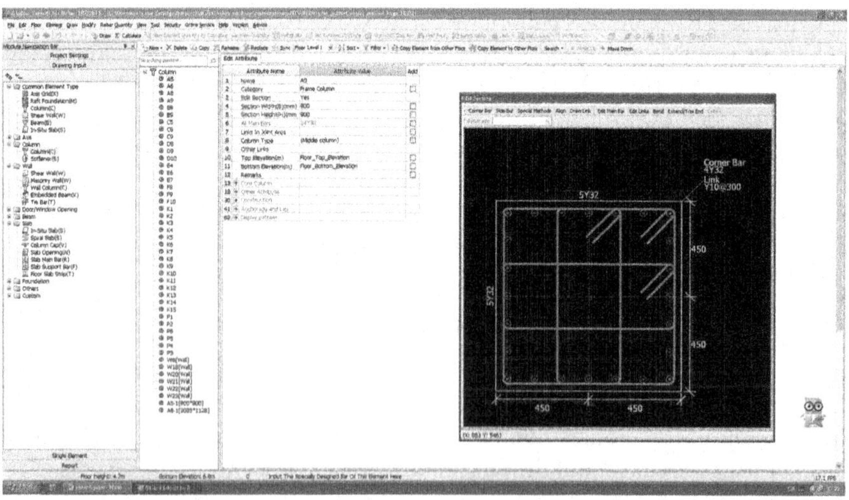

Figure 5.10 The arrangement and manual amendment of rebar in a column.

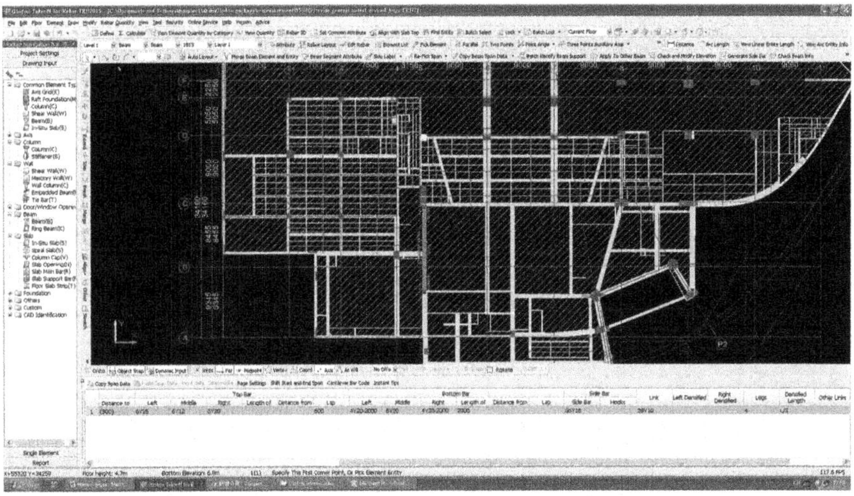

Figure 5.11 The arrangement and manual amendment of rebar in a beam.

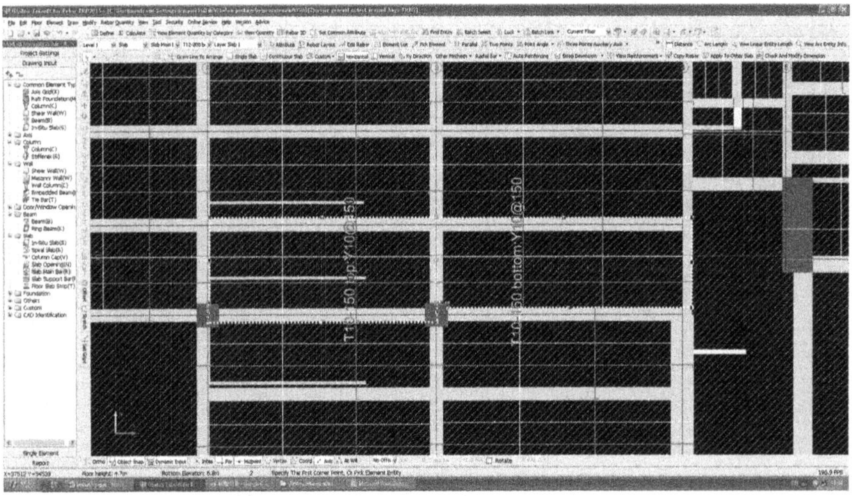

Figure 5.12 The arrangement and manual amendment of rebar in a slab.

demonstration of the rebar configurations, and integrating measurement rules for QTO as its calculation basis. Figures 5.10–5.12 present the manual amendment of some typical structural components (i.e., columns, beams, and slabs).

5.3.3.2 *QTO result presentation*

After developing the QS-BIM model, QS extracted and presented quantity information in Excel files for subsequent final account settlement. The Excel files included the breakdown of element, as well as summaries for different levels

Floor: Level 1 (Drawing Input)					Total Weight: 218410.152 Kg				
Mark	Type	Size	Rebar Shape	Formula	No. of Bars	Total No.	Length of Each Bar (m)	Total Length (m)	Total Weight (kg)
Element Name: 1B1				Element Qty: 1	Bar Weight: 185.967Kg				
Element Location: <8-1638,B+2103><P2-2310,B+2103>									
1	Y	20	300 7047 300	300+312-30+312+5783+300+350-30+350	2	2	7.647	15.294	37.721
2	Y	20	300 7047 300	300+312-30+312+5783+300+350-30+350	2	2	7.647	15.294	37.721
3	Y	20	6925	240+312+5783+240+350	2	2	6.925	13.85	34.16
4	Y	20	6925	240+312+5783+240+350	2	2	6.925	13.85	34.16
5	Y	10	440 340	2*((400-2*30)+(500-2*30))+(36.5*d)	20	20	1.925	38.5	23.735
6	Y	10	440 127	2*(((400-2*30-20)/3*1+20)+(500-2*30))+(36.5*d)	20	20	1.498	29.96	18.47
Element Name: 1B12				Element Qty: 1	Bar Weight: 746.622Kg				
Element Location: <5+4348,C-250><5+1864,D-271>									
1	Y	12	180 9666 283	180+259-30+259+9053+408	2	2	10.129	20.258	17.985
2	Y	12	180 9666 283	180+259-30+259+9053+408	1	1	10.129	10.129	8.993
3	Y	16	9370 67	12*d+9053+155-30+67	6	6	9.437	56.622	89.378
4	Y	40	251 9666 433	480+259+9053+480+78	2	2	10.35	20.7	204.222
5	Y	40	251 9666 433	480+259+9053+480+78	1	1	10.35	10.35	102.111
6	Y	40	7525	-(0.1*9317)+259+9053-(0.1*9317)+77	3	3	7.526	22.578	222.75
7	Y	10	740 340	2*((400-2*30)+(800-2*30))+(36.5*d)	65	65	2.525	164.125	101.183

Figure 5.13 The presentation of rebar remeasurement in Cubicost TRB.

and the overall project. These Excel files clearly organised the quantity information and, most importantly, demonstrated the calculation process of each rebar quantity. At this stage, Cubicost TRB structured the extracted quantities according to the diameters and types of rebar. Besides, the QTO generated by Cubicost TRB also presented the rebar shapes, calculation process, formula, and calculated quantities clearly (see Figure 5.13). However, manual amendment was necessary to further rearrange the table according to client's requirements. In addition, QS had to prepare another format of rebar similar to that of the traditional method so that contractors could compare and agree on each cost item.

5.3.3.3 Settling the final account

After preparing the rebar remeasurement documents, the responsibilities of QS shifted into reaching an agreement with contractors to finalise and settle down the project account. For this project, this was done by a cost review meeting between QS and contractors after the construction work completed. Before the meeting started, QS and contractors exchanged their assessment concerning the total construction work. Specifically, the assessment of QS was based on the drawings for construction, prepared by contractors before construction commenced. QS then verified the quantities and quotations submitted by contractors, compared contractor's assessment with their own remeasurement results, and marked any suspicious items and comments for discussion at the meeting.

During the meeting, the BIM model displaying detailed rebar information greatly facilitated negotiation. It provided a vivid demonstration of the rebar layout and made it possible for QS and contractors to clearly understand the design and construction details. Based on the model and assessment, they exchanged comments and, for the most part, reached an agreement regarding the quantities and rates of each proposed cost items. After the meeting, QS prepared a recommended final account based on the meeting minutes. A report of non-agreed items was also included in the final account portfolio, listing the contractor's and QS' assessed quantities/rates and their reasoning for the assessment. After

Table 5.5 BIM-based and manual methods for rebar measurement

Item	Manual	Glodon
1.Column		
a. Main bars		✓
b. Links		✓
c. Starter bars		
2.Beam		
a. Main bars	Continuous beam more than 3 rebars on the side/sloping beams	Good for curved beams
b. Stirrup	>20 legs	
c. Side bars		✓
3.Wall		
a. Main bars		✓
b. Links		✓ (new version)
4.Slab		
a. Main bars		✓
5.Water tanks	✓	Can use single element but not efficient
6.Staircase	✓	Can use single element but not efficient
7.Other(e.g. Tie bars)	✓	Can use single element but not efficient

Note: tick representing more suitable

deliberating the QS' recommendation on the final account settlement, the client ultimately decided to close out the Final Account and issue the Final Certificate.

5.3.4 Findings and lesson learnt

Case No.3 presents the implementation of BIM for remeasurement of rebar during the final account stage. In this case project, QS determined the modelling methods for different components according to the efficiency and accuracy to develop the rebar model in BIM. For example, an automated method was adopted for rebar measurement of typical structural components (e.g., slabs, beams, walls, and columns). This implied that the modelling and QTO of these components largely relied on BIM with little manual amendment effort. For those complicated structural elements, such as water tanks, staircases, and tie bars, BIM proves less efficient. More manual effort must be invested to model the rebar and extract the quantities.

Table 5.5 summarises the suitability of BIM based on the manual remeasurement method for various rebar types at different locations. It is identified that BIM is useful in measuring typical structural components in which the binders and additional rebar can be edited freely whilst the horizontal and vertical rebar can be arranged systematically for walls and slabs. Beams, comparatively, require more time and effort to lay out the rebar in Cubicost TRB if stirrups are involved. The information on top/bottom/side bars, as well as legs/stirrup of beam, could be inputted into Cubicost TRB. However, when the level of the beam is not the same (e.g., stepped beam), modelling may be more difficult. It would be easier for the stepped part to be measured manually. Slab rebar can be separated into top and bottom rebar and can be adjusted according to the shape of the slab.

The setting of rebar to fit both the General Notes and British Standard proved the most difficult part for software targeting reinforcement. Features like drawing numbers and shape codes had to be incorporated into the schedule for easier comparison of quantities. The setting in Cubicost TRB can cater to most situations, but QS can still input the quantities into the formula manually when necessary.

5.4 Summary

This chapter presented three case studies relating to BIM and QS practices with a focus on automated QTO, tender document preparation, and material remeasurement for final accounts. Following the QS-BIM execution plan described in Section 4.2, this chapter elaborated the software solutions, practice changes, benefits, and difficulties of BIM implementation for QS based on real-life cases. Traditional PDF or CAD drawings were prevailing compelling QS to develop QS-BIMs, as well as enrich them with the appropriate information. The extra efforts turned out to be worthwhile in these three projects. BIM helped QS and other stakeholders in calculation, automation, communication, visualisation, and negotiation. Most stakeholders deemed the use of BIM valuable, but prudently maintained that risk management needed to be considered before its implementation could be scaled up within their respective company. These cases offer far from holistic BIM experiences. BIM roll-out faces a learning curve when it comes to exploring organisational structures, resource allocation, and changing process and protocol.

6 Big data for construction cost management

QS professionals for years have been somewhat constrained to work within the narrow or incomplete data provided by designers and other stakeholders. Whilst the 'big data' movement has burgeoned in various fields, such as biology, medicine, finance, and public governance, its application in QS has been somewhat lacklustre. Surveyors, business leaders, professional bodies, and the like question whether the QS profession can join the movement and harness the power of big data. The quick answer is 'yes'. QS firms already possess a large amount of quality data that could, if properly channelled, ease the systemic QS problems presented in Section 1.4. This chapter explores the potential of big data for construction cost management. After outlining what constitutes big data, it provides some theoretical explanations for the big data phenomenon, citing Herbert A. Simon's bounded rationality theory amongst others. After presenting several cases of big data in connection with construction cost management, the chapter reviews the prospects and challenges of applying innovative big data methods to such an arguably conventional profession and posits BIM as one possible information platform through which to collect, store, and subsequently utilise big data for AEC practices.

6.1 What is big data?

Although there is little consensus on what big data really is, big data can be considered as "a collection of data sets so large and complex that it becomes difficult to process using the available database management tools". As compared to separate smaller sets with the same total amount of data, big data allows such correlations to be found to spot business trends, determine the quality of research, prevent diseases, link legal citations, combat crimes, and determine real-time roadway traffic conditions. Padhy (2013) suggested that big data is a collection of data sets so large and complicated that it becomes difficult to process using traditional data management tools. Likewise, Mayer-Schönberger and Cukier (2013) propose big data as things one can do at a large scale that cannot be done at a smaller scale to create a new form of value in living, working, science, and industry by changing markets, organisations, relationship between people, and more.

Notwithstanding the disagreement on terminology, researchers tend to adopt Gartner's three defining characteristics of big data, i.e. volume, variety, and velocity, otherwise known as the three 'Vs' (Zikopoulos and Eaton, 2011; McAfee et al., 2012). Volume is the quantities of records, transactions, tables, or files; velocity finds expression in batch, near time, real time and streams; and variety can be structured, unstructured, semi-structured, and a combination thereof (Russom, 2011; Zaslavsky et al., 2013). It was predicted that data will grow exponentially; every millisecond, second, minute, hour, day, week, month, year, data is relentlessly generated from web logs, sensors networks, unstructured social networking data, and streamed video and audio. Big data analytics analyse various types of colossal amounts of data to uncover hidden patterns, unknown correlations, and other useful information, to provide better business forecasting and decision-making (Zhu et al., 2016), or in other words, to create the fourth 'V', i.e. value (Figure 6.1).

Big data has rapidly become the 'heart of the talk' of this era across a wide array of areas (Raj et al., 2015). It is transforming various research disciplines, including biology (Howe et al., 2008); medical science (Murdoch and Detsky, 2013); ecological science (Hampton et al., 2013); business (Chen et al., 2013), urban planning (Hao et al., '015); public governance (Misuraca et al., 2014); and innovation, competition, and productivity (Gobble, 2013). Government agencies use big data to generate statistics, which help them understand local and global patterns and trends, and ultimately improve their services. Its ability to harness information in novel ways to create insights and services makes big data a crucial source of innovation (Mayer-Schönberger and Cukier, 2013).

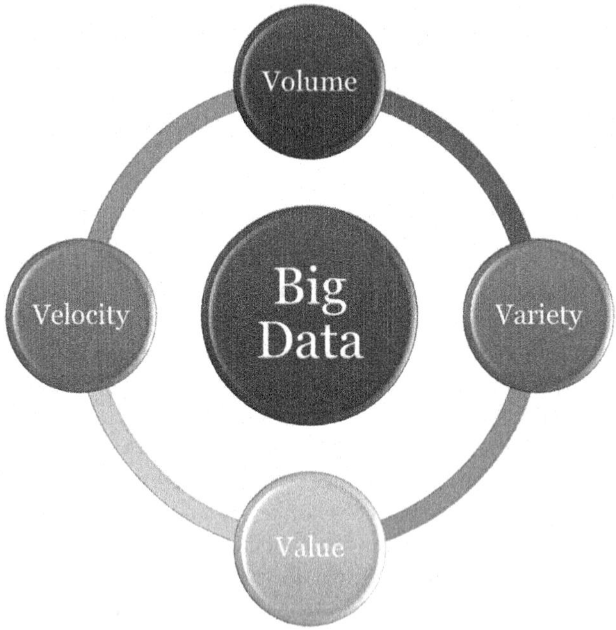

Figure 6.1 Big data with 4 Vs

By analysing big data, researchers aim to identify some 'latent knowledge' (Agrawal, 2006) or 'actionable information' (World Economic Forum, 2012) to inform future decision-making. Likewise, UPMC believes that by enhancing the collection of big data, and translating it into actionable insights, big data analytics can drive greater efficiency, and help the personalisation of healthcare, and better define patient populations with a greater level of granularity. The McKinsey Global Institute (MGI) asserts that data is becoming a factor of production, like physical or human capital. In 2010, it was estimated that this industry on its own was worth over US$100 billion and growing at about 10% a year (Weng and Weng, 2013). Another source (NewsOn6.com) estimated that the global big data market will reach $118.52 billion by 2022, rising to a compound annual growth rate of 26% from 2014–2022 (NewsOn6, 2018).

6.2 Why is big data in vogue?

A hype of big data is somewhat arising from every corner of the world. A plethora of bestsellers and online articles eloquently promote the use of big data. But what theoretical perspectives best explain this big data phenomenon? Researchers and business leaders need to understand the reasons behind before they jump into the sea of big data.

6.2.1 *Data, information, and knowledge*

A good point of departure would be to understand the relationship between data, information, and knowledge. There is a common view that data is raw numbers and facts, information is processed data incorporated with meaningful structures, and knowledge is authenticated information (see Figure 6.2. Vance, 1997; Dretske, 1981; Machlup, 1983). This view imagines that data is a prerequisite for information, and information is a prerequisite for knowledge. Reversely, Tuomi (1999) argued data emerges from information, which emerges from existing knowledge. This notion denies the existence of pure 'raw data' but believes that all data is generated by the intervention of human thinking. Alavi and Leidner (2001) more specifically defined knowledge as information processed by the human mind as facts, concepts, ideas, judgments, and interpretations of information.

Particularly, knowledge is perceived or created by processing data and information incorporated with human thinking. Knowledge can be viewed as a capability which possibly has impact on future actions (Carlsson et al., 1996), and the capacity to use information, such as interpreting information and selecting useful information for decision-making (Watson, 1999). Although the hierarchy of data, information, and knowledge cannot be clearly defined, the purpose of collecting data is to extract useful information and create knowledge so as to support informed decisions.

Organisations tend to have a formal process for digesting and prioritising information in order to inform decision-making. By amassing and examining data, they convert the results into practical information. Poleto et al. (2015) envisaged this progression as transpiring as and within a kind of massive, "generalist

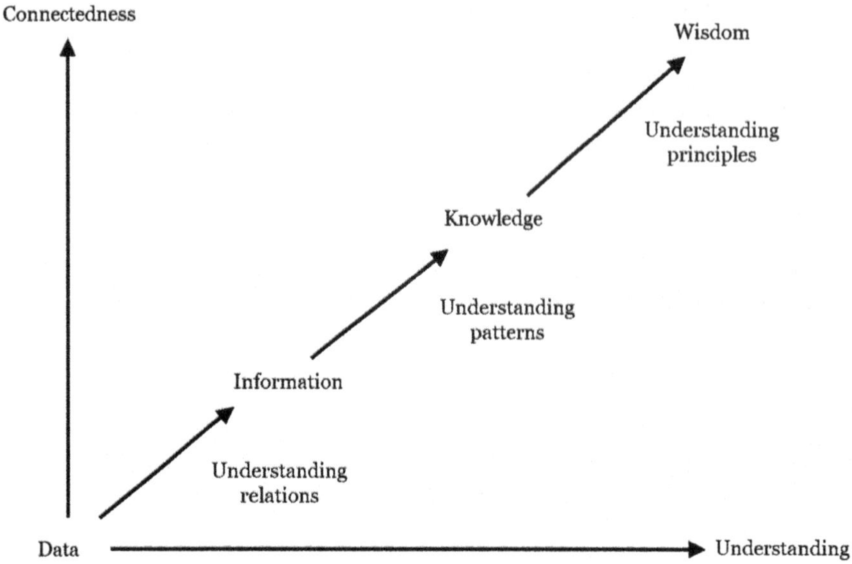

Figure 6.2 Data, information, knowledge, and wisdom.

repository, a processor core with the appropriate intelligence and user-friendly interface. The repository must be filled with data originating from many different kinds of external and internal data sources". He equated big data as data warehouses or data marts. Big data is popular because it has benefits, including the power of prediction, and improved decision-making.

6.2.2 Unbinding the 'bounded rationality'

Following the thread that big data allows more informed decision-making, decision science poses a sensible theoretical perspective to explain the craze of big data. Accordingly, AEC activities, including construction cost management, involve a large number of decisions made. This resonated with Flanagan and Lu (2008), who argued that managing a construction project, certainly including its cost management, exploits available information and knowledge to make a web of decisions across the processes of architecture, engineering, and construction. From a decision science perspective, management essentially is making a web of decisions based on the information and knowledge available (Grant, 1996). Instead of pursuing completely correct decision-making, decision scientists nowadays seem to be advocating informed decisions (Bekker et al., 1999). A decision-maker should be well informed of the facts, implications, and consequences when making a rational decision (Amendola, 2002), even though the outcome may prove other than what was expected. The main objective of managing data and information is to support decision-making.

Simon's (1982) bounded rationality assumes that human decision-makers work under three constraints: (1) only limited information is available regarding possible alternatives and their consequences, (2) human mind has only limited capacity to process the available information, and (3) only a limited amount of time is allowed for decision-making (see Figure 6.3). Big data could ease these restrictions to allow more reliable decisions to be made.

First, the large volume of big data can ease the limitation of the information available for decision-making. Decision-makers normally have limited information to allow them to make informed decisions. Decision-makers often rely on small data that is carefully curated using traditional methods such as sampling, experiments, or other ethnographic methods. Big data can alleviate the potential bias inherent in small data and provide a fuller picture to have a closer claim of objective truth. The variety of big data further enhances the quality information that is available to decision-makers. It can measure something from different perspectives and thus can allow decision-makers to have a closer access to the totality of a subject under consideration. The big volume of data with different varieties can also allow hidden patterns, unknown correlations, and other useful information to be uncovered.

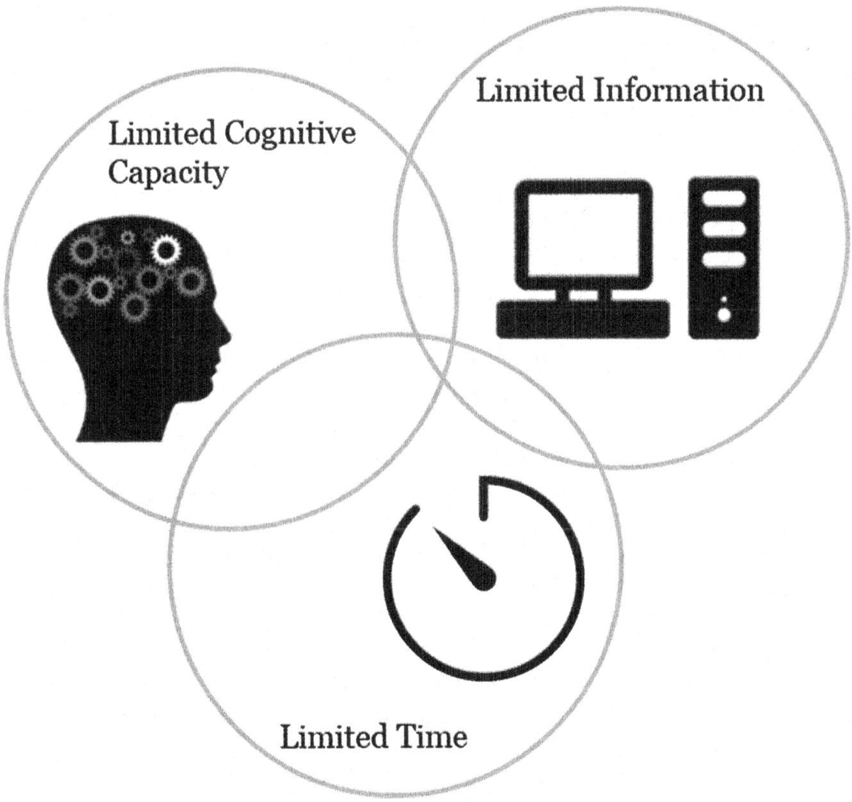

Figure 6.3 Illustration of bounded rationality.

Second, the velocity of big data provides both challenges and opportunities to deal with the last two limitations as articulated in Simon's bounded rationality. The large volume of big data coming in a high speed further challenges human mind's cognitive capacity to process the data. It must engage advanced information and communication technologies (ICTs). As pointed out by Padhy (2013), even using traditional data management tools showed their incapability. The rapidly incoming big data with velocity further squeezes the limited amount of time allowed for decision-makers. For example, in social media, every day, millions of photos, tweets, and videos are uploaded on Facebook, Twitter, and YouTube. Big data analysts need to timely handle and analyse the data to derive useful information, for example, to understand and respond to emergencies that have been covered in the big social media data. Big data analytics provide real-time insights, which need to be actioned quickly to gain better value. Organisations that employ swift decision-making processes are able to realise better value from real-time data analytics and build a strong competitive advantage.

In summary, AEC activities, including construction cost management, can be perceived as making an array of decision-making *per se* based on available information and knowledge. Big data, with its characteristics of volume, variety, and velocity, provides an opportunity to help make informed decision. It can help increase our rationality which is bounded by limited information, limited cognitive capability, and limited time we have to make a decision. That can partially explain why the world is enthusiastically promoting and developing big data and its analytics.

6.3 Cases of big data for construction cost management

The immediate questions are 'Do we QS have big data?', and 'How exactly can we unleash the power of big data for cost management?'. The answer to the first question is 'yes'. Unlike the stereotype that the industry overall is 'small' in data generation, construction, including cost management, is actually 'data rich' by nature. Cost management data, such as architecture and engineering drawings, BoQ, bidding documents, schedule of rates, process/routing sheets, master production schedules, accounting records, supplier and catalogue information, labour rates and standard time data, and repair and maintenance schedules, is restlessly generating from multiple projects throughout their life cycle ranging from inception, design, construction through to operation. QS play an important role during this process. When QS business is done, big data, with volume, velocity, and variety, is unintentionally left over in the company.

The data are voluminous. A single BoQ is as thick as a book, let alone other data as mentioned above. The data volume might not be as big as tera- or zettabytes as normally seen in social media hubs, but the strength of the QS data is it being well structured and thus may contain more meaningful information than unstructured data with the same volume. The data is of variety. It could be graphics (e.g., drawings), texts (e.g., BoQ, schedule of rates, memos, emails), voice (e.g., phone calls), pictures (e.g., progress report photos), or videos (e.g., webcams).

It could be paper-based as seen in the old days, or in digital format created by spreadsheets, AutoCAD, Office Automation software, laser scanning, RFID, or other ICTs more recently. The data is incoming with velocity. Every day, every minute, batches of data are generated from your employees, from the sites, and from the intensive communication amongst them. Some of the data lost, some of the data, however, is well captured in QS firms' internal management information systems. The big data is similar to the 'buried asset' that can be harnessed to further enhance QS business and overcome the problems relating to the profession, for example, tedious and error-prone measurement, inaccurate estimate, and time pressure to deliver a QS service.

6.3.1 Big data to help prepare tendering and cost estimate

In order to put this into context, take a real-life Company, but with the pseudonym of Ace. Ace is a surveying firm based in Hong Kong, where QS is far from the sunset profession as many worry surveying is becoming. In fact, one of the most successful surveyors had become the Chief Executive of Hong Kong, as a Special Administrative Region (SAR) of China. Ace has been involved in QS business in Hong Kong for more than 30 years, and in recent years, its business has expanded to Mainland China to embrace the booming construction market there. The company has mainly been serving as a consultant to its clients, ranging from private to public clients in projects of all sizes. The main business is thus similar to the ones as described in Chapter 1, for example, hired by a designer, or a client directly, to provide QS-related services such as preliminary cost estimate, design-stage cost plan, tendering, cost control, and variations and final account (see Figure 1.1).

For every project it serves, Ace kept the drawings provided by the designers, BoQ they created, bidding documents and schedule of rates provided by prospective bidders, normally 3–5 bidders per project, master production schedules, accounting records, supplier and catalogue information, labour rates and standard time data, claims and communication, interim payment records, and so on. In addition to the standard documents (e.g., BoQ) which must be in a standardised format and handed over to various parties, the company uniquely enjoys a software system developed in-house to record all the spreadsheets and the BoQ that were generated by the system. The system was developed in the 1980s, from today's perspective a bit outdated, clumsy, and difficult to interoperate with other software, with the exception of specific spreadsheets, and connect to the Internet. However, the end result is the access to big data left over unintentionally, which can be utilised to enhance its QS service.

Company Ace can use the big data to develop a comprehensive in-house 'code book' that can be applied to almost every step of QS service to enhance its accuracy and productivity. For example, taking off quantities from detailed design drawings is relatively straightforward. However, as mentioned previously in this book, some special finishes, trades, or necessary deduction may not appear in every project. It is expected that the big data, containing the finishes, trades, and deduction practices in previous projects, could be formally analysed and recorded in the code book for future uses. QS, in particular those new to the company, can reuse the information without having to develop everything from scratch or consult experienced colleagues, who are understandably also quite busy. Unit rates are equally important. By analysing the pricing items submitted by the bidders, together with appropriate consideration of the market conditions and project characteristics, the pricing information can actually serve as a more accurate reference than small data in providing pre-contract estimates and cost plans. This echoes the LLN; the average of the results obtained from a large number of trials should converge to a certain number or range. The big data can provide more accurate information as it covers a large number of past items.

By reusing the quantity and price information derived from the big data, QS can provide quicker services, such as QTO or pre-tender estimate, as some of the rare or non-standard items are readily coded in the code book. This is about the traditional QS business, but big data can relieve the time pressure of QS. As observed, the AEC industry moves quickly. QS can also provide reasonably more accurate estimation during the design stage cost plan or even earlier, during feasibility cost estimation. Rather than simply being a subcontractor of the designer, QS entered the upstream business by directly serving as the financial consultant to clients, in particular high-end clients without construction knowledge. Company Ace also feels that the image of the profession has changed.

What Ace did was to hire a big data analyst, who worked with the directors to seek their support, with the company's in-house IT team to have the data interoperability, and equally importantly with the frontline surveyors to understand their needs. The database of the in-house system was opened for analysis. Some of the paper documents need to be digitalised and made machine-readable using optical character recognition (OCR) and other technologies. Data cleansing helped detect and correct incomplete, incorrect, inaccurate, or irrelevant parts of the raw data. The big data was analysed using regular statistical software such as SPSS or R without necessarily involving fascinating terms such as 'pattern finding algorithms', 'unattended machine learning', 'deep learning', or 'artificial intelligence'. Data visualisation is also important, as it can attract employees' attention and further seek their true engagement in the big data movement.

6.3.2 *Big data to help prepare bidding*

A company called Bauhinia is used as an example. Bauhinia is a construction company, often acting as a main contractor in bidding and undertaking construction business. We could use above Company Ace in this context, but normally a construction company has its own in-house QS team. Company Ace, serving as the consultant to a client or a designer, normally sits the other side of the table when work with the construction company. When the tender documents including 'instructions to tenders', 'preliminaries specification', 'technical specification', 'pricing documents' (e.g., BoQ, schedule of rates), and 'drawings' of a project are realised, Bauhinia decided to bid for it. The construction company needs to prepare bidding with its QS team. As mentioned previously, nowadays, the time left for contractors to prepare bidding is exceedingly short.

Normally, the QS team will engage its speciality suppliers and subcontractors. Understandably, the time left for them to prepare bidding is even shorter, as the QS team will check the individual bids and assemble them in a final bid. Everyone on the bidder side is under time pressure to prepare bidding with shorter time but with more precision. It is particularly important to issue an accurate estimate, or it will cause huge loss in later stages after the project contract is awarded. There is a 'winner's curse', wherein the winner will tend to overpay due to emotional reasons or incomplete information. Big data can help in bidding, one of the most critical decisions made by a construction business.

Company Bauhinia over the past years has accumulated huge volume of data when bidding and undertaking construction projects. What it did was similar to Company Ace using the big data to develop a comprehensive in-house 'code book' containing rich quantity and price information. By comparing the tendering documents with the in-house code book, it can quickly reflect any suspect design errors or incomplete information included in the BoQ. The latest quantity and price information summarised from multiple trades in multiple projects truly reflects the company's operation and management capability. It can be used to triangulate the bids supplied by subcontractors and adjust the final bid for tendering the project. The readily available code book can certainly expedite the bidding process, and the information derived from the big data, in line with the LLN, is more accurate to reach a reasonably precise bid, so as to reduce the risks in later stages once the project is awarded and carried out.

Like Ace, Bauhinia also hired a big data analyst, who worked with the directors to seek their support, with the company's in-house IT team to have the data interoperability, and equally importantly with the frontline surveyors to understand their needs. The database of the in-house system was circulated for analysis. It is important to follow the clients' standard preliminaries or specific standards clauses, if there is, in developing the in-house code book, normally it should be consistent with client's cost items and trades. Professional bodies such as the RICS are helping in aligning these to enhance interoperability. Data cleansing helped detect and correct incomplete, incorrect, inaccurate, or irrelevant parts of the raw data. Regular statistical software, such as SPSS or R, scrutinised the big data. Data visualisation is important since it can attract interest from the users so that they are willing to further participate in the big data movement.

6.3.3 *Big data to help analyse bidders' behaviour*

Checking tenders and issuing a tender report is a significant part of a QS' job. It is not as simple as performing the tasks including arithmetical check, consistency of pricing, opening two-envelope tenders, choosing the lowest bidder, and making a recommendation concerning the appointment of a contractor to carry out the work. There are many speculative bidders that should be recognised and separated from those bona fide bidders. Some bids are exceedingly low. Some bids are suicidal and non-serious. Some companies' offers are consistently low. Some bids are marked up, whilst others are purposely marked down. Some lowest price is simply because of arithmetical mistakes, which can explain the 'winners' curse'. All these risks should be properly identified; otherwise, they will cause huge loss to project stakeholders, for example, cost overrun or claims, in later stages.

Big data, if it has captured all the bidding documents from a large number of bidders, can help in analysing their behaviour. The bidding documents actually contain a lot of useful information that can be mined from them. A single bid from one contractor may not tell much but comparing all the bids against itself longitudinally or against other bidders horizontally will tell a lot. The surveying firm Ace noticed that one of the biggest contractors has over the years consistently offered a low price, which seems too good to be true. However, by mining its bidding documents, it is discovered that the company has a stable logistics and supply chain that can also offer lower price. The bids are serious, and the company is a competitive contractor. By mining the documents, Ace also noticed that a company offers reasonably low price to win the contracts, but the company is notorious in claiming later, as a way to make up for its initially low price. These are useful hidden patterns or unknown correlations that would not be possible to be uncovered using small data.

Big data can also be used in other QS businesses, such as cost control, variations, and final accounts. For example, big data can help recognise the critical points for variations; therefore, special cost control attentions can be paid to the points. Big data can also help identify the claims and make sensible responses. It has also been reported that big data can be used to identify frauds in construction (e.g., bid rigging, by analysing bidders' social networks and bidding patterns). Not all these big data cases are fully elaborated here.

6.4 BIM and big data

Big data emerging from the AEC industry does not have to be linked to BIM, but the introduction of BIM as the 'disrupter' of the industry provides a unique opportunity to exploit big data. Properly accrued, BIM and big data can allow a virtuous circle in enhancing each other to facilitate construction cost management. As mentioned previously, BIM thrives due to the value of its contained information. BIM is not static, but continuously enriched with real-time information incoming as the systems operate. Data comes into a BIM-centred system with a certain volume, variety, and velocity. For example, a large volume of traffic data relating to an urban area generates each second, and ideally, that data should be captured by information technologies affiliated to BIM for cost applications. The increasing affordability of sensor networks and computation powers makes real-time collection of big volume of data possible. Researchers are also exploring advanced technologies such as laser scanning or photogrammetry to collect the real-time data of a construction site and synchronise it with the cyber system to derive a so-called 'as-built' BIM (Chen et al., 2015). By combining BIM and advanced ICTs, one can make more informed decisions on cost estimating and planning control for project delivery. It would be a misnomer to emphasise that today's cost estimating only occurs during the early stages of projects. The dynamic information feedback or feedforward to support decisions during a project life cycle do not follow a traditional one-way mechanism.

BIM models are important components of big data for VDC, any decision of which is unavoidably associated with construction cost. BIM can be further integrated with a large variety of project information including the geometric, semantic, and topological information (Schlueter and Thesseling, 2009). With the fast development of information technology and management, the new tools under the concept of BIM have become popular amongst designers and engineers. BIM is fast becoming the most common tool for collecting, storing, and managing proactive big data. BIM models are mostly well-structured big data and allow users to query them, but gradually become semi-structured owing to their combination with other information systems or sensing technologies, such as GIS, laser scanning, and RFID (Isikdag et al., 2007; Meadati et al., 2010; Xue et al., 2018b). These applications stimulate the exponential increase of data amount and complexity, leading to the big data in the construction industry.

A prevailing misconception surrounding BIM and big data is that they are automatically available as a panacea, able to solve any problem in cost management. Actually, data and BIM do not transpire spontaneously and have to be linked up and constantly updated. At this juncture, it would be good to revisit the cost data scheme as shown in Figure 2.3 and further connected to the QS-BIM as discussed in Chapter 4. The item codes in Figure 2.3 offer the keys to link the cost data scheme and the code books as mentioned above in Section 5.3. The more accurate cost data can be periodically updated from big data analytics, the cost data, ideally in Extensible Markup Language (XML), is better placed in a middle-ware so that they can be linked to various BIM software packages to have a maximum level of interoperability.

6.5 Prospects and challenges of big data for construction cost management

The strengths of big data lie in its volume, velocity, and variety that can be harnessed to derive useful information to support decision-making. However, they are not without their challenges in application to construction cost management.

6.5.1 *Big data technologies*

The fast-growing information and computing capability has brought about a large number of technologies available for big data collection. Big data is available everywhere in and out of the life cycle of construction projects. Web-downloading data is an important and easily accessible part of the reactive data. It may include legislation and policy documents and data generated from government practice, which may involve the financial incentives for the references of cost management. However, some organisations unintentionally maintained data related to construction cost, such as order information, suppliers' information, and construction personnel salaries, which also serve as the reactive data as they are not proactively collected. However, the data is seldom used for deeply identifying patterns; instead, they are just kept in stakeholders' computers. The collection method of the public data is usually downloading from relevant websites as they are public, whilst the project-related data in organisations cannot be easily obtained as they belong to different data owners. Not especially advanced technologies are required for extracting value from reactive data; however, proactive data collection may involve several technologies.

Proactive big data can be geometric and spatial data for a large number of scenarios, such as process tracking (El-Omari and Moselhi, 2011; Turkan et al., 2012), site monitoring (Teizer et al., 2007), safety management (Cheng and Teizer, 2013), and FM (Shen et al., 2012). The real-time decision-making is normally based on real-time data collection with these techniques sought for swift responses. Currently, the proactive big data is generated from data collection techniques, such as laser scanning, RFID, augmented reality (AR), GIS, global positioning systems (GPS), and sensors (Chen et al., 2015). In the life cycle of construction projects, there are many scenarios with the potential for data acquisition and decision-making. Different scenarios use different measures and techniques for data collection. The proactively collected data after leveraged for achieving the designed purposes may become the reactive data for other usages. Data sources in the life cycle are massive and complicated, comprising big data for providing a great deal of information for decision-making in cost management.

6.5.2 *Life cycle costing (LCC) enabled by BIM and big data*

Big data together with BIM can be used to present the building information, such as the building structure, construction schedule, material use, construction cost, labour behaviours, and energy and water use throughout the construction project life cycle. It boasts significant advantages over the building life cycle in design, construction, and FM (Yan and Demian, 2008), such as time saving, human resource saving, cost saving, quality improvement, sustainability, and creativity, as BIM-based design will continue its benefits to the following phases (i.e. construction, operation, and demolition). The purpose of using BIM is to provide more information to reach the objectives concerning buildings more efficiently. For example, the use of cloud-based BIM has been regarded as an approach for providing big data for the management decisions of building sustainability in the whole life cycle (Wong and Zhou, 2015).

Life cycle cost (LCC) is concerned with optimising VFM in the ownership of physical assets by taking into consideration all the cost factors relating to the asset during its operational life (Woodward, 1997). Owners, users, and managers need to make decisions on the acquisition and ongoing use of many different assets including items of equipment and the facilities to house them. The initial capital outlay cost is usually clearly defined and is often a key factor influencing the choice of assets given a number of alternatives from which to select. Consideration of the costs over the whole life of an asset provides a sound basis for decision-making. There are many tasks in the LCC, which requires information of a large volume and high quality for cost estimation. Typical cost estimates for a system may include acquisition costs (or design and development costs), operating costs (or cost of failures), cost of repairs, cost for spares, downtime costs, and loss of production, cost of preventive maintenance, cost for predictive maintenance, disposal costs, as well as other accounting or financial elements. Big data, including BIM and other data collected, in the construction project life cycle can provide the LCC with abundant information if the data collection is well designed in the life cycle scenario.

BIM and big data can fully prepare all the required cost values as inputs; then, the cost analysts can seamlessly present a complete cost analysis for individual tasks in a spreadsheet. Big data and BIM join the fragmented information on construction cost in the life cycle and provide a full picture for decision-makers' life cycle considerations. Some research has attempted to integrate the information of BIM and big data to develop more information-abundant LCC tools. For example, Kehily and Underwood (2017) demonstrated the BIM-based LCC solution, where LCC integrates into the BIM process by embedding an LCC calculation model structure within an existing BIM technology (Kehily and Underwood, 2017). This is realised through information exchange between the BIM and LCC tool and improving cost databases, simplifying the data access as well as storage possibilities to ease and extend the use of LCC.

BIM and big data both serve as platforms for enabling the data acquisition and information flow throughout all stages. When all practitioners have access to the same information, and the possibility to transfer information from the model to the different calculation and analysis tools, decision-making in all scenarios can be based on the up-to-date assessment. With more information provided at the early phases, more analysis may be conducted to make changes as early as possible to reduce construction and operation costs. Life cycle cost analysis may occur throughout each stage by using information in the BIM or information from different cost databases. Results from the analysis are then fed back to the BIM and databases for use in other analysis, decision-making or for later use in FM during the operation phase.

6.5.3 *Breaking down the silos*

Currently, many big data sets, including cost data like estimates, BoQ, interim payment records, and claim cases, unintentionally remain when a project concludes. The amassed data can be a corporate asset, the mining of which allows companies to make better business predictions and decisions, and ultimately, enhance their productivity and sharpen their competitive advantage. The big data, an untapped goldmine, naturally created from business processes tends to be considered better than experimental data or simulation data as it potentially contains more ground truth with respect to social reality than traditional instruments. Big data portrays a fuller picture of a subject matter, which permits a stronger claim to objective truth. It can therefore be predicted that data owners will become more protective of their big data and reluctant to share their corporate assets with others, even though their value has yet to be fully appreciated.

The AEC industry should abandon its silo mentality and share big data for cost management. Interestingly, although Hong Kong is one of the most commercial places in the world and safeguards trade interests, the SAR's QS industry has actually benefited from open big data (e.g., the tender price index published by RLB (https://goo.gl/HJUr64), Construction Cost Handbook by Arcadis (https://goo.gl/9yGm6E), or the SMM4 jointly published by HKIS). Associated organisations and various government departments have disseminated data in a way that strikes a balance between protecting private commercial benefits and achieving public good.

6.5.4 *Big data ethics*

A double-edged sword, the capture and use of big data has both benefits and risks. Ever since the advent of big data, there has been concern over the ethical ramifications of data analysts misusing its power. For example, the big data in Company Ace can help analyse the bidders' behaviour as naturally left over in the big data sets. Nevertheless, whether it is ethical to analyse the data and make use of the big data analytical results is open to vigorous disputes. Traditional de-identification approaches (e.g., anonymisation, pseudonymisation, or encryption) to protect privacy and confidentiality and allow analysis to proceed are now problematic in big data, as it is the power of big data analytics that even anonymised data can be re-identified and attributed to specific individuals (Ohm, 2009). The ownership, liability, and subsequent use of the data has not been a problem before but now triggered a series of ethical issues, none of which is resolvable with an easy answer.

6.6 Summary

The construction industry has vacillated in the face of the global big data movement. Owing to the temporary nature of projects and AEC's prevailing fragmentation and discontinuity, the industry still depends on conventional practices based on 'small data'. This chapter endeavoured to demystify the concept of big data by reviewing its definitions and theories and explain its popularity. Its major appeal concerns better support of decision-making than small data. This chapter confirmed that construction, including cost management, is 'data rich' by nature. Many QS firms or organisations have accumulated very reliable big data. This 'buried asset' can be harnessed to enhance QS business. A 'source of truth' in a collaborative working environment, the introduction of BIM provides an unprecedented opportunity to collect, store, and exploit the value of big data. The characteristics of big data, including its volume, velocity, and variety, can overcome the QS systemic profession's problems (e.g., tedious and error-prone measurement, inaccurate estimation, fragmentation, and time pressure to deliver QS services). This chapter provided three examples of how BIM can help by preparing tendering and cost estimation, preparing bidding, and analysing bidders' behaviour, whilst wider applications wait for both practitioners and researchers to explore and pursue. Finally, this chapter discussed the prospects and challenges of big data for construction cost management, such as changing the mentality around sharing big data and the ethics of doing so.

7 Current challenges and future outlooks

It took a decade for people to realise that adoption of BIM for construction, including its cost management, is not as straightforward as imagined. Despite the great progress achieved, there are still many challenges. This chapter provides an overview of the current challenges and future outlooks associated with BIM implementation for cost management. It covers the challenges organised under six categories, namely, standards, technical, economic, organisational, legal and contractual, and cultural challenges. These challenges are not mutually exclusive with each other but closely intertwined. Instead of ranking these challenges from least to most important, this chapter attempts to elaborate them as in depth as possible and permit readers to personally determine their magnitudes. Despite the challenges, readers and prospective BIM users should not be overwhelmed or feel pessimistic about the future of BIM for QS. There is no perfect solution or quick fix to these difficulties, but they are manageable and a bright future exists for the application of BIM in the construction industry.

7.1 Lack of standards

Many industries operate based on standards, but construction is a particular one that relies on standards to govern its procedures, activities, and deliverables. The first challenge in using BIM for QS concerns the lack of relevant standards regarding how BIM and cost estimation can operate together (Wu et al., 2014). No comprehensive and explicit written rules exist to inform stakeholders what information and in what LoD the information must be presented within the model in order to support cost estimation. No significant case studies exist from which to draw on as a benchmark (Gu and London, 2010). Designers and contractors, for example, frequently asked the format of the BIM files and in what media (e.g., hard disk, memory stick, or CD) when handing in the BIM files. Governments, for equal opportunities for all the software vendors, normally will not specify a file format. Without a clear standard or guideline BIM modellers often feel obliged to under or over-inform, which exacerbates the risk of design errors in a BIM. Errors and misinformation complicate a QS capacity to manage and register the required materials within a model for the development of cost estimates (Wu et al., 2014). Enormous efforts are squandered to filter, request, or supplement incomplete and inaccurate information.

In addition, even if BIM software can automatically generate a project's QTO, these take-offs rarely come in a format suitable for pricing. Matipa et al. (2010) suggested that current measurement standards were developed for conventional paper-based surveying method. However, BIM is incompatible with current global measurement standards (e.g., the NZS 4202:1995 Standard Method of Measurement of Building Works; the Hong Kong Standard Method of Measurement of Building Works). Such incongruity prevents QS firms from using BIM for cost estimation and BoQ generation.

The lack of standards for coding objects within BIM models also hinders the application of BIM for QS (Smith, 2016). Coding standards for BIM-based QS are essential as they help convert the information in the building information models into BoQ. If designers can better understand how QS categorise and measure by following a coding standard, the likelihood the BIM model they produce can be directly used by QS for measurement rockets. One of the early efforts to propose protocols for coding objects emanated from RICS in the UK. RICS has been collaborating with industry representatives to develop new rules of measurement and a concise, comprehensive BIM for QS specification report. Although a release date has yet to be issued, this book's authors anticipate a list of clearly defined information which suits the particular requirements of business and QS. Via these standardised measurement rules and coding protocols, BIM can be better compliant with the QS measurement standards.

7.2 Technical challenges

Despite the impressive progress in BIM technology development, technical challenges still mount. The technical challenges faced by the implementation of BIM generally come from the following aspects. First, fuelled by the fragmented nature of the construction industry, vendors often run software in proprietary data file formats. Such fragmentation results from the use of many different types of BIM software as their interoperability and compatibility is unsatisfactory. Matipa et al. (2008) identified several challenges associated with interoperability, including the difficulty of translating basic geometry information, lack of information in some file formats, and the complexities in translating semantic information between various file formats. The lack of interoperability is a serious barrier restricting the implementation of BIM for QS (Olatunji, 2011). Different BIM software, with various model building up methods, lead to different quantity measurement calculations. Sabol (2008), Steel et al. (2012), and Hardin and McCool (2015) also argued that many BIM-based estimation applications currently do not accommodate bidirectional information exchange. For example, if the designer uses Graphisoft ArchiCAD for BIM model generation, but QS adopts Autodesk Revit for QS, information loss will occur when QS receives the BIM model from the designer, and the corresponding items will be missed when estimating and scheduling quantities.

The current BIM software may allow the quantity information contained in the model to be constantly transferred and updated during design changes, but not the cost information (Kymmell, 2008). Therefore, bidirectional information exchange is necessary to keep the BIM models updated and increase information integration for many value-adding applications including the QS (Sabol, 2008). In an attempt to ensure bidirectional exchange and overcome the data interoperability issue, open data standards, such as IFC (buildingSMART, 2016) discussed in Section 2.2.3, have been applied to the construction industry. By creating a neutral format, IFC can help govern the data exchange amongst CAD software, BIM software, estimation software tools, and other construction management software. Nevertheless, there are still several compatibility issues associated with IFC (Stanley and Thurnell, 2014). Information, even in the IFC format, can still be overlooked or wrongly represented during the exchange between different software.

Secondly, BIM-based estimation software applications remain immature. The construction industry has realised relatively low capacity, capability, and development extent of BIM software. Many BIM software applications are developed to meet the requirements of a specific purpose or specialty. They are not advanced enough to substitute QS expert knowledge, particularly in dealing with some detailed aspects. Therefore, existing BIM solutions are not completely trustworthy as QS tools. For instance, an application developed for the designer to construct the conceptual estimation is of little use to QS in order to generate the BoQ. Although commercial BIM software applications (e.g., Autodesk Revit) provide a set of API for the BIM users to develop their own add-ons, not all QS firms have their own in-house software development teams to formulate the required add-ons for QS. Additionally, these BIM software applications often rely on a single or couple external cost databases. QS has to make a difficult choice between adopting the BIM-based estimation software application compatible with the company legacy of cost database and adopting the application that could best fit their needs (Forgues and Iordanova, 2010; Forgues et al., 2012).

Thirdly, cybersecurity also poses a practical challenge for BIM advancement (Mahamadu et al., 2013; Solihin and Eastman, 2015). Cybersecurity refers to

> The approach and actions associated with security risk management processes followed by organisations and states to protect confidentiality, integrity, and availability of data and assets used in cyber space. The concept includes guidelines, policies, and collections of safeguards, technologies, tools, and training to provide the best protection for the state of the cyber environment and its users.
>
> (Schatz et al., 2017)

When making the BIM model accessible to different stakeholders through the Internet, cybersecurity can be a serious concern because of the possibility of online unauthorised access and copyright infringement (Chien et al., 2014). Cybersecurity can be conducted in several ways from controlling physical access to the hardware to protecting from malicious viruses, hackers, or other threats. However, few project teams using BIM on a daily basis enjoy the luxury of sufficient information technology and support from experts with good insight into the nature of construction management (Holzer, 2015). The lack of in-house expertise makes cybersecurity extremely difficult.

7.3 Economic challenges

Although several academic studies and industrial reports have demonstrated the evidence of the benefits of BIM implementation, mainly anecdotal (e.g., Azhar, 2011; Giel and Issa, 2011; Barlish and Sullivan, 2012; Neelamkavil and Ahamed, 2012), the economic issues associated with it need to be further addressed. Some of these issues can be easily quantified in monetary terms, whilst others are more nuanced.

A firm's initial investment in BIM essentially constitutes the cost of the software, hardware, and training. Given small to medium-sized enterprises' (SMEs) limited financial capacity to invest in emerging digital technologies and capabilities, bigger companies tend to be the ones subscribing to comprehensive BIM services. SMEs often regard the initial investment of BIM programming as too expensive and prioritise other expenditures over it (Goucher and Thurairajah, 2013). Apart from the initial investment, looming software and hardware upgrades add to SMEs' disinclination. In addition, the development of a BIM model incurs cost and work hours as the complexity of the project mandates a greater LoD. Another source of cost, one difficult to quantify, relates to the potential reduction in staff members' productivity as they face the BIM learning curve.

The business value of any innovative process or system shoulders uncertain outcomes (Bryde et al., 2013). Zhou et al. (2012) surveyed a wide range of AEC stakeholders irrespective of their company size. Half of their interviewees professed that for them the BIM benefits did not compensate for the capital costs. Smith's (2014) subsequent study found similar results. However, all interviewees of that same study agreed the 'wait-and-see' approach no longer works for the firms that want to be key industry players, and yet they still expressed concern about potential losses caused by the use of BIM. This suggests the fear of squandering time and funds on software, hardware, and labour training outweighing the fear of failing to maintain a competitive advantage by not using BIM.

Unfortunately, the vast majority of construction companies often find themselves trying to secure enough business to meet costs rather than in a position to explore new technology, processes, and businesses models. SMEs choose to wait in the wings to see if their competitors using BIM enjoy higher a rate of selection or commercial return over time. This leads to BIM being more commonly used in large-scale projects for bigger clients with the wherewithal to demand all project participants employ BIM and its compatible software (Smith, 2014). It is critical to make a strong business case for the long-run development of BIM in construction cost management. Current misconceptions that BIM is too arduous to make a stable return against its cost can be associated with the challenges above (e.g., lack of standards or technical hurdles). Without standards, BIM users like QS and developers can only use it in pilot projects instead of applying BIM widely to achieve a 'scale of economy'. Without cost codes and handy BIM libraries in place, BIM users have to develop them from scratch, and such developed libraries cannot be shared across the entire industry, which unavoidably increases the unit cost of developing the BIM.

7.4 Organisational challenges

Deployment and implementation of BIM to a certain degree changes the operational process and organisational structure of how a construction project transpires. BIM is not merely about a set of technologies, but also a mediator of innovation that can re-engineer ineffective processes and develop new business strategies (Dainty et al., 2007). Senior management should support new processes and strategies for effective implementation. Nevertheless, Elmualim and Gilder (2014) found three major organisational challenges of BIM implementation, namely, training staff the new process, effectively implementing the new process and workflow, and achieving sufficient understanding of BIM. Lu and Korman (2010), Aibinu and Venkatesh (2013), and many other researchers identified the greatest challenge of using BIM in construction projects as the implementation process itself.

Organisational challenges are seen with the same companies and their intra-firm project organisations (Winch, 1989). Implementing BIM will unavoidably change reporting structures, communication patterns, and regular working processes; therefore, the resistance to change is considerably overwhelming. Anything relating to organisational change, particularly in construction, is not easy. Management of change associated with the implementation of BIM is important (Bryde et al., 2013). Some companies, when trying to use BIM in their projects, may set up a BIM group and let BIM guys work in parallel with the project team (Nagalingam et al., 2013).

Organisational challenges are also found in the collaboration amongst various entities involved in the same project (Lu and Korman, 2010; Wang et al., 2013), which is called inter-firm project organisations (Winch, 1989). A typical construction project necessitates cooperation and information exchange amongst a variety of parties, such as the client, designer, engineers, QS, and contractors. Fully IPD with multidisciplinary project teams working on a single integrated and compatible BIM model is essential for the optimal use of BIM. Currently, the scope for integration remains limited (Smith, 2014). It is very difficult, if not entirely impossible, for a traditional procurement model like the dominant model, particularly in public works, DBB. In a DBB project, the expert knowledge on cost provided by a QS consultant or on buildability provided by an experienced contractor is not available at the early design stage, as the two stakeholders have yet to be appointed. This explains advocating IPD to promote BIM adoption. Chapter 8 discusses procurement models and organisational changes in greater detail.

Even though it is possible to assemble all the different parties under the same roof of an inter-firm project organisation, each party can implement their own BIM systems. Each company normally has its own preferred software applications for design and analysis. It is very rare that a sole technology is used on a single project by all parties and across all phases of the project life cycle. Rather than dependent on a single model, project team members typically rely on a number of purpose-built models. However, as suggested by Olatunji et al. (2010), the use of BIM for QS requires the collaboration, database integration, and commitment of firms to BIM. Purpose-built, fragmented BIM models thus significantly limit the effectiveness of BIM-based QS.

7.5 Legal and contractual challenges

Other than finance and organisational issues, QS firms will experience legal and contractual issues associated with BIM implementation. The use of BIM encourages seamless collaboration between all stakeholders, which may, however, blur the level of responsibility amongst individual stakeholders and the assignment of liability amongst them. Several legal issues should be considered, which include but are not limited to the ownership of the BIM model, the use and distribution of that model, and IP rights (Lu and Korman, 2010). In other words, it is important to determine who has the right to proprietorship and access to the valuable information contained in the BIM model, who is responsible for charging and updating the information in the model, and what should be done if errors surface in the model.

The problem of BIM model ownership, as well as the data contained in the model, becomes extremely serious in the situation of the joint authorship of BIM models. In addition, the accuracy of the BIM information entails substantial risk since inaccurate information can easily lead to poor decision making (Ghaffarianhoseini et al., 2017). Without a clear identification of legal responsibility for errors and problems within a BIM, stakeholders suffer from uncertainty and spend additional time and money on clash detection and correction (Smith, 2014).

Smith (2014) noted that all legal and contractual issues relating to BIM implementation should be answered in order to ensure the effectiveness and efficiency of BIM implementation and achieve the full collaborative potential of BIM. Azhar (2011) suggested that a BIM manager and relevant professionals be assigned to track and control errors and make decisions regarding responsibilities. Stakeholders should reach an agreement on BIM model access, information security and documentation, and transmitting issues early on in a project life cycle.

A new form of contract could be implemented to avoid arguments concerning BIM responsibilities and liabilities. Alternatively, the protocols of BIM that cover all legal issues should be created to allow stakeholders to retain a contract type which they are accustomed to (Azhar, 2011). However, extra attention should be paid to conflict between BIM protocols and the principal contract. For example, the BIM protocol may require a more comprehensive IP licensing procedure than that provided in the principal contract. The administration of information accuracy within a BIM model will cause additional costs (Ghaffarianhoseini et al., 2017).

7.6 Cultural challenges

Last but not the least, cultural challenges affect the implementation of BIM for QS. The construction industry has been labelled as a very old-fashioned industry, and some researchers deem cultural intractability a much greater hindrance than any relating to BIM's technological difficulties (Stanley and Thurnell, 2014). Most AEC practitioners began their careers when such techniques were the only techniques, and even now only a few surveying degree programmes offer extensive coursework in BIM. They favour the conventional paper-based QS methods that have been utilised for decades and are reluctant to change. Changing their minds and asking them to use BIM thus is extremely difficult.

Apart from the reluctance to change, another cultural challenge is that the staff members in the same office could have vastly different understandings of BIM programs and applications (Fox and Hietanen, 2007). Many firms see this disparity as the main BIM inhibitor (Vass and Gustavsson, 2017). Under the pressure of tight schedules, it is risky and hardly possible to catch up staff members' understanding of BIM for BIM-based QS. Such disparity makes it difficult to harvest the full benefits of BIM. In this circumstance, firms could end up feeling dissatisfied with the outcomes and switch back to conventional paper-based methods afterwards.

7.7 Summary

BIM adoption for construction cost management faces many challenges. First, it lacks agreeable and readily operable standards for BIM to be applied to this rather traditional profession, despite the fact that various general BIM standards have been instituted in several major economies. Technological challenges like rapid BIM development, information standards (e.g., LoD, IFC), interoperability, and cybersecurity exist. Despite widespread anecdotal evidence of BIM's benefits, the industry still craves a stronger business case to justify the wider adoption of BIM for QS work, especially for SMEs. BIM does not operate in a vacuum. The success or failure of BIM for QS relies on the amenable organisational support, yet current practices are not equipped enough to support this either due to their fear of change or the dominant nature of current procurement models, which separate design and build activities. The chapter went on to explore the legal, contractual, and cultural challenges, which have long plagued the industry and its adoption of BIM for QS. Although discussed blow by blow, the challenges in actuality all intertwine with one another. For example, some economic challenges such as the cost to develop a QS-BIM would lower if modelling technologies could make faster use of big data. Prospective QS using BIM should not be disheartened by these challenges. In the long run, the industry will more than likely appreciate extensive adoption of BIM and big data for construction cost management.

8 Good practices for adopting BIM for cost management

This chapter recommends the good practices of adopting BIM for cost management. The authors pose seven, not necessarily 'best', but 'good' practices based on their analysis of QS, BIM, big data, and their prospects and challenges. They range from fostering research and development (R&D), pre-graduation and post-professional training, cost benefit sharing, and innovative procedures to BIM localisation. Adopting BIM for QS does not have to be a holistic and in one go. Rather, companies can use BIM to solve their most urgent issues or in promising areas before scaling up BIM to all their QS work. Some of the practices desire the efforts from QS firms, some from professional bodies, and others from all stakeholders, who should work together to achieve the goal of making BIM an ally of QS.

8.1 Encouraging research and development (R&D)

Continuous R&D is an important strategy for companies to enhance their competitiveness by developing new services or improving existing services to their clients. As BIM and other relevant technologies are evolving rapidly, innovation is needed to advance the application of BIM in real-world projects (Cheng and Lu, 2015). In-house R&D is therefore indispensable in order to continuously improve companies' knowledge and competency to provide improved QS and cost management services.

Companies can explore new possibilities of in-house R&D by integrating cutting-edge technologies, such as VR, AR, mixed reality (MR), and 3D scanning with BIM and QS. These new technologies allow the synchronisation between the physical and cyber worlds, which provides metadata and more accurate information for BIM use and QS. Moreover, the large amount of information and data generated can be stored in the cloud for easy storage, inquiry, and retrieval. Afterwards, machine learning (ML) and deep learning (DL) (i.e., a hierarchy of ML approaches using visual, audio, digital, and other forms of recognition and processing) can be utilised to mine the big data of BIM and QS spreadsheets for more hidden information, patterns, and knowledge of construction industry and QS. For example, by mining and learning the data of all the QS data in the database of a company, the factors that affect projects costs can be detected and an empirical formula extracted. This empirical formula can help approximate new projects' costs. With enough data sample and training, the empirical formula will be increasingly accurate until QS will only need to input the factors and the computer can automatically derive an exact cost based on the formula. Besides, companies can collaborate with universities or other research institutions to stimulate more innovative ideas and applications. For example, the University of Salford and Beijing Construction Engineering Group (BCEGI) conducted a partnership project to transform BCEGI's traditional workflow towards a BIM-based innovative collaborative work practice and process (Alazmeh et al., 2017).

8.2 Continuous training and education

BIM like any emerging technology comes with a learning curve (Lu et al., 2013). Construction industry practitioners require training and practical experience before they can operate BIM software with dexterity. Training and education ensure the implementation of state-of-the-art technologies in QS. As mentioned in Section 7, the lack of knowledgeable practitioners impedes widespread BIM adoption. Therefore, sufficient training and education should be provided to help staff members better understand BIM and the basic operations of its software applications (Thurairajah and Goucher, 2013). Individualised training programmes together with a qualification framework for different BIM roles (e.g., BIM managers, BIM modellers, and BIM analysts) focused on practice will quickly raise the BIM competence of a company. The training programmes can be designed based on BIM protocols, which according to AIA Document E203 (AIA, 2013b) and UK Construction Industry Council BIM protocol (UKCIC, 2013) should include the following issues:

1 Definitions and terms (i.e., words and phrases used in the protocol);
2 Coordination and resolution of conflicts (e.g., possible conflicts and corresponding coordination strategies, and priority of contract documents);
3 Model ownership;
4 Model requirements (e.g., model standards, file formats and electronic data exchange, and model usage);
5 Model management (e.g., the requirements for managing the model, initial responsibilities and ongoing responsibilities of the employer and the project team member, liability in respect of a model, model archives, and terminations); and
6 Security remedies (i.e., security requirements and provisions, such as secured log-in with access privileges, hierarchical model administration structure, check-in/check-out and version lock, and technical provisions; Singh et al., 2011).

Different types of training materials and approaches can be adopted in conjunction including, but not limited to:

- Traditional training approaches (e.g., help menus, FAQs, and helpdesk);
- Open-source training materials (e.g., technical support blogs, online webpages, and online communication community);

- A shared folder to share tool usage guidance;
- Interactive tutorials by experienced staff or experts; and
- An expert directory maintained in the tool-compatibility matrix (Singh et al., 2011).

The trained staff members should participate in actual projects where BIM is used to delivery QS. In this way, their knowledge of BIM can be reinforced. Besides, knowledge sharing programmes should also be added to the training system. Knowledge sharing includes regular meetings to review and discuss BIM best practices and lessons learnt, industry conferences and webinars, as well as knowledge exchange between professionals with deep design, construction, and/or management experience with less experienced but BIM-capable professionals in order to blend expertise and develop holistic understanding of the processes, benefits, and challenges from traditional to BIM-empowered practice (Hong Kong Construction Industry Council [HKCIC], 2015a).

Attention should also be paid to continuing professional development (CPD), which regularly evaluates staff BIM proficiency and conformance to standard BIM processes to ensure staff keep their skills and knowledge up to date and remain competent to perform their duties. CPD splits into formal CPD and informal CPD. The former includes structured learning with clear learning objectives and outcomes, such as a professional course, structured online training, and technical authorship, whilst the latter constitutes any self-managed learning related to one's profession (RICS, 2016). RICS requires CPD identify experienced staff that may serve as BIM leaders or knowledge resources that maintain active memberships in industry organisations and attend industry events.

Moreover, education fundamentally supplies and enriches talent in the construction industry. Both universities and professional organisations like RICS, AIA, and HKCIC offer excellent educational materials. Collaboration with universities to develop BIM training for students and young professionals is also strongly recommended. Universities help lay the theoretical foundation, and industry gives students the chance to practice what they have learnt, whilst professional organisations, which combine practical experience and expertise, offer quality continuing education for current staff.

8.3 Making a strong BIM business case

Although it is common for top executives of AEC firms buy into the latest tech fads without clearly comprehending their economic returns, BIM needs to tender a genuine economic foundation in order to sustain in the business world. Without evident benefits, companies may not invest in BIM. Its implementation is costly (e.g., purchase and maintenance of hardware and software, and cost of training staff) and must compete for companies' limited financial resources. To that end, a strong value-driven business case must be made for BIM and big data in application to QS.

QS and cost estimation pose processes that a BIM model can inform and to some extent perform (e.g., automatic or semi-automatic generation of an accurate QTO) early in the design process and adapt to additions and modifications potentially saving time and money, as well as avoiding budget overruns. It also provides a platform for designers to evaluate the cost effects of their design changes in a timely manner, which can help prevent excessive cost overruns due to design changes (CICRG, 2010). With model-based estimating software, design authoring software, and cost data, BIM offers many potential benefits in QS including:

- Precise estimation of material quantities and quick revisions;
- Budget control within constraints by frequent preliminary cost estimation throughout the design progresses;
- Better visual representation of project and construction elements that need to be taken off and priced;
- Cost information for the project owner during early decision-making phases;
- Focus on more value-adding activities in estimation, such as generating pricing and factoring risks;
- Exploration of different design options and concepts given the owner's budget;
- Saved QS time by allowing QS to focus on more important issues; and
- Quick determination of specific objects' costs.

Regardless of these benefits, what practitioners care more about is the real costs to implement BIM. It was reported that less than 20% of surveyed private sectors used BIM for model-based QTO to enhance estimation process (HKCIC, 2015b). There are numerous issues including lack of standards, technical, organisational, and other challenges as elaborated in Chapter 7, which may not allow benefits to offset the investment in many BIM-enabled QS practices. To make a value-driven business case, a company should first of all revisit its expertise, vision, and value so that the mangers can make a determination on the trade-off between value and information (e.g., LoD) based on a CBA. The BIM infrastructure, together with the processes and protocols, needs to be reviewed and upgraded periodically in line with the latest technology and market developments.

8.4 Sharing costs and benefits

Previous studies have focused on finding BIM a business case or finding empirical evidence of BIM's benefits for QS. Even as the above section suggested, BIM can make a strong business case, it is recommendable to further explore cost-benefit sharing, with a view to sustaining BIM in QS business. Costs and benefits mean different things to different project participants, which are largely different profit centres inherently safeguarding their own interests. Each stakeholder expects some benefits from BIM implementation and is thus prepared to incur some obligations to obtain them. The increase of overall benefits of the project does not necessarily mean a benefit increase to an individual party involved. When there is a misalignment between the overall and own interests, it is unavoidable that some participants will act in his/her own best interests, which is at odds with the spirit of BIM.

The industry needs to find a proper cost and benefit sharing mechanism to encourage different stakeholders to collaborate consummately based on BIM. Researchers such as Chang (2014) from University College London (UCL) and Lu et al. (2017) from the University of Hong Kong have started to look at this cost/benefit sharing issue. It needs to recognise the economic legitimacy of BIM participants, in addition to the good wish of 'being collaborative', and tries to cost-benefit rebalance model, with a view to incentivising project participants to consummately implement BIM and harness its full power. Traditional reporting structure amongst clients, designers, consultants (including QS), and contractors needs to be reviewed with the intervention of BIM-enabled practices; so does the fee structure governing the stakeholders.

8.5 Embracing innovative procurement models

Although BIM and QS can work along in traditional DBB procurement model or the like, the full potential of BIM can only be realised wherein all the stakeholders are collaborative consummately. In spite of their advantages, traditional procurement models have long been criticised for the deep-rooted flaw of exaggerating the fragmentation and discontinuity of the construction industry (Masterman, 2003). A dialogue between stakeholders involved in traditional procurement models is often difficult due to that stakeholders tend to get appointed sequentially with little opportunity for engagement when it counts most for life cycle BIM. There has been an evolution of procurement models with an effort towards emphasising stakeholder collaborations, rather than hard power such as process (Rowlinson and Mcdermott, 1999). BIM has the soft power to solve the intrinsic problems existing in traditional procurement models, primarily due to that BIM allows the integration of 'back-office' systems with those used for transactions, and the accurate transmission and processing of information. Some typical ingredients of BIM-based traditional procurement include the use of BIM as template for generating tender documents, the cooperation between the designer and the contractor, early involvement of contractors, life cycle BIM implementations, and risk sharing of collaborating parties. There thus emerges a call for advanced procurement models to truly unleash the benefits made possible by BIM.

Notably, the IPD method is deemed as an innovative procurement model that can alleviate some deep-rooted problems of traditional procurement strategies. Over recent years, it has been strongly promoted by the AIA, with its official working definition released in 2007 (AIA, 2007a). Unlike the traditional procurement methods that create isolated islands amongst project participants, IPD endeavours to bridge the islands with sharing risks, knowledge, and responsibilities. The IPD approach concentrates on the integration of people, systems, business structures, and practices into a process that collaboratively harnesses the knowledge, talents, and insights of all project stakeholders to increase project value, reduce waste, and optimise efficiency through all phases of design, fabrication, and construction (AIA, 2007b). It attempts to share the risks, responsibilities, and liabilities for project delivery fairly amongst the main project participants. The core project team members in an IPD project, usually the owner, the designer, and the contractor, stipulate their responsibilities and commitment to each other in a multi-party agreement or equal interlocking agreements. In this regard, contractors can contribute their expertise of construction techniques early in the design phase to improve the constructability and, therefore, the project quality and financial performance (AIA, 2007a).

As IPD and BIM share similar principles of collaboration, the fusion of BIM and IPD is gradually changing the working patterns of building projects and is expected to solve the well-known problems in terms of fragmentation and discontinuity of the industry. Indeed, IPD and BIM are tightly bound with each other towards successful collaboration in construction project delivery. The combination of IPD and BIM can also bring some benefits. For example, by integrating contractors early in the planning phase, the constructability of the project can be largely increased with fewer problems in the construction phase. Besides, using integrated expertise, the design scheme could be in higher quality, having the potential to cutting waste and energy consumption. It can be expected that IPD, together with BIM, will change the organisational structure and, thus, enable the main principles and benefits of IPD to be implemented.

8.6 BIM localisation

One good practice that is recommendable for adopting BIM for QS practice is to 'localise' BIM. Current BIM technologies, organisational settings, and related working protocols are developed from a 'standard' background without necessarily considering the local needs. Local specifications play an important role in the domain of cost management. They not only regulate the format and contents of cost reports; more importantly, they help advise the general procedure for construction cost estimation and conduct other cost-related activities based on local requirements. QS should rely on local specifications as they conduct their various cost-related activities (e.g., cost estimation and construction tender document requirements) with client expectations. For example, it has been reflected that existing BIM solutions cannot cover the MEP systems and trades that are unique in a region, for example, Hong Kong or Mainland China. It is onerous for users to further develop the standard objects to suit their needs. They also need to develop their local cost data and the measurement methods (e.g., deduction of some intersecting sections of beams and columns, or their formwork and falsework).

BIM localisation is a process to analyse, adapt, and adopt general BIM knowledge (i.e., BIM technology, process, and policy) to suit the specific needs (e.g., regulations, building codes, and cultures) of a local context. There are three dimensions of BIM to be localised, including technology, process, and policy. Here, technology not only includes BIM data, software, and hardware provided by vendors, but also integral platforms to support effective communication and collaboration amongst different stakeholders (Succar, 2009; Jung and Joo, 2011). BIM process denotes a series of ordering work activities of BIM creation, management, and utilisation to support the project tasks (Succar, 2009; Wong et al., 2010). The localisation of BIM process concerns the organisational and managerial issues resulting from integrating BIM and existing AEC practices, especially when organisations are developing their BIM strategies based on global best practices or BIM execution plans originated from other countries. BIM policy can be regarded as a course or principle of action adopted or proposed by a government, party, business, or individual as a fundamental reference to guide the BIM-based project delivery (Succar, 2009; Jung and Joo, 2011).

BIM localisation can be easily linked to real-life situations of using BIM for business. For example, when BIM is implemented for construction cost management in China or Hong Kong, the 'standard' BIM solutions may not provide the applicable cost codes or handy cost elements as China uses Guobiao (GB) system and Hong Kong adopts a British Standard (BS). An individual firm may have to adapt the standard BIM library and naming system to its local demand. This seems trivial, but when it comes to real-life sizeable projects, the task could be burdensome. The BIM execution plan is another example. It as a company-level policy may function well in one company (e.g., in the US), or a specific business (e.g., MEP design) may not do the same in another company (e.g., in China), or another business (e.g., QS). Great efforts should be input to localise the general BIM knowledge, including technology, process, and policy, to make it a true and handy ally.

BIM software vendors present mixed feelings in joining this BIM localisation. Some are hesitating, as they need to differentiate various niche markets in their product development, which seems against the 'scale of economy'. However, to have localised products will encourage high BIM adoption. That may ultimately increase the BIM business. Companies like Glodon are working closely with practitioners in various markets to localise their products, for example, to develop the trades and cost data that are more suitable for Mainland China's norm system or Hong Kong's SMM.

8.7 Think big, act small

In the BIM discourse, a holistic BIM can be understood as that all AEC activities are based on BIM from the very beginning and the BIM is updated promptly by all professionals and in turn supports their works throughout the project life cycle (Coates et al., 2010). Whilst it is true that BIM can realise its full potential if it can be holistically implemented in AEC projects, a strategy advisable to executives of QS firms is that they do not have to achieve the holistic BIM in one go. Rather, they can implement BIM to solve their most urgent and promising issues, such as automated QTO, preparing for tendering, or forming the cost control baseline. In doing so, employees can be encouraged by the small but achievable benefits and begin to embrace BIM consummately. It has been noticed that some companies implemented BIM in a smaller scale and in some specific aspects of their AEC business. After accumulating sufficient experiences, they started to incrementally scale up to the whole company and extend to the more aspects of their AEC business. Drastic changes such as business process reengineering (BPR) for BIM are not always realistic. Sometimes, they present more risks to the smooth operation of the companies.

8.8 Summary

In view of the challenges as elaborated in Chapter 7, this chapter recommended the good practices to address the challenges and for more active adoption of BIM for cost management. There are many technical challenges and widespread lack of BIM knowledge. It is therefore recommended to boost continuous R&D to develop the BIM knowledge and competency of companies in the long run. It is understood that there is a learning curve for BIM adoption for construction cost management; therefore, continuing professional development or curriculum and pedagogical development are encouraged. BIM should truly bring benefits to QS firms so it can sustain in the business world. It is really desirable to have a strong business case for BIM. In addition to making a BIM surplus, it is also critical to redistribute the costs and benefits amongst the stakeholders including QS to incentivise them to embrace BIM consummately. That is the spirit to unleash the power of BIM. The success of BIM cannot be achieved without the support from its related organisation structures. BIM for QS needs a more collaborative environment endorsed by innovative, integrated procurement models. IPD is a promising way to reduce the fragmentation and discontinuity that can subsequently foster a collaborative environment for BIM adoption. One good practice that is recommendable for adopting BIM for QS practice is BIM localisation amidst the global trend of BIM internationalisation and standardisation. BIM localisation is a process to analyse, adapt, and adopt general BIM knowledge to suit the specific needs of a local context. Adopting BIM for QS does not have to be a holistic BIM in one go; rather, companies can use BIM to solve their most urgent issues or in promising areas before BIM is scaled up to all their QS businesses.

9 The future of BIM and big data in quantity surveying

This chapter discusses the prospects of BIM and big data in QS. A clear trend is that BIM is increasingly migrating from traditional local personal computer (PC) basis to become cloud BIM. Stakeholders can work on BIM remotely from their portable devices without having to be nailed down together. They can also enjoy ubiquitous BIM services. Whenever they need information for decision-making (e.g., project progress, interim payment), they can access it. In this sense, BIM is realising its essential meaning as a digital representation of physical and functional characteristics of a facility to support decisions during its life cycle. With that, data is accumulated from cloud sourcing and stays in the BIM to form the big data to further support decision-making. With the advancement of BIM technologies and big data, a computational BIM can be expected to provide real-time quantity information (e.g., GFA, cost, or materials) for designers and contractors to make more informed decisions. QS can be freed from laborious, low-tech measurement works and become consultants to advice on macro or more strategic cost issues. QS profession can expand itself to the upstream AEC businesses.

9.1 Cloud BIM

Cloud BIM is a rapidly growing industry trend, which allows the sharing of BIM in a more efficient manner. Citing from the NIST definition, Mell and Grance (2011) classify cloud computing as

> Enabling ubiquitous, convenient, on-demand network access to a shared pool of configurable computing resources (e.g., networks, servers, storages, applications and services) that can be rapidly provisioned and released with minimal management effort or service provider interaction.

Examples of shared BIM resources range from components and libraries to applications and services. They can be rapidly provided and released with minimal management effort or service provider interaction. BIM can be imported into a cloud platform for remote access via various communication devices (e.g., PCs, tablets, and smart phones) (Wong et al., 2014a).

Cloud-based BIM serves more than just design phase 3D modelling and visualisation. As an extensive life cycle data management platform, BIM facilitates the administration and analysis of multiple dimensions (e.g., time, space, cost, sustainability, and FM) (Chien et al., 2017). As the stakeholders and professionals involved in a construction project tend to be geographically dispersed, it can also enable communication and collaboration through remote access whilst displaying the most up-to-date model and information.

Similar to the three categories of cloud computing (Kang et al., 2008), cloud BIM can also be separated into three categories, namely, private cloud BIM, public cloud BIM, and mixed cloud BIM. A private cloud BIM is exclusively owned and used by a single organisation. Only the staff of the organisation and its contracted customer can access the confidential cloud BIM. The private cloud BIM may rely on commercial cloud platforms, such as Autodesk BIM 360 and the Graphisoft BIM server. However, if the client calls for a stricter confidentiality or the commercial platform cannot cater to the information processing, private cloud BIM can also be developed in-house. This occurs by exporting the BIM data into an open format (e.g., IFC or gbXML) and then rendering the data using interactive online 3D graphics (e.g., Web Graphics Library [WebGL]).

A public cloud BIM provides a free cloud access to the general population. Users waive their intellectual technology and proprietary rights in the spirit of the open-source movement and propagation of BIM. Another key reason for public cloud BIM involves the establishment of an open online BIM library, through which the public can download and utilise the library's BIM elements, as well as enrich the families and elements by uploading their own designs. Public cloud BIM can also be utilised by industry communities or associations to align local standards or assist SMEs in BIM applications. Such industry communities can publish standard BIM object parameters, LOD, and working or collaboration procedures on a public cloud BIM platform. Organisations and professionals with access to the public cloud BIM platform can refer to the standards for individual operation and partnerships.

A mixed cloud BIM is a hybrid of private and public cloud BIM systems. On the one hand, BIM projects can be shared amongst involved parties in-house. On the other hand, the BIM elements contained in the project can be shared to a designated group of stakeholders or open to the public. The integration of private and public cloud provides various business opportunities for different organisations as these two types of cloud BIM can be combined in various ways (Wong et al., 2014). Organisations can choose the right combination that suits their needs when moving to the trend of cloud BIM.

9.2 Ubiquitous BIM service

For the cloud BIM practice, BIM supports model design and viewing whilst the cloud integrates the data storage, sharing, and networking. When BIM is coupled with the virtual and mobile features of the cloud, models can be worked on simultaneously by various professions, clashes can be identified quickly, and the most updated progress of project can be accessed by anyone in any location (even on the construction site) at any time. Another advantage of cloud BIM that might appeal to the BIM users is the potential reduction in investment for BIM hardware and in-house system development. When moving from stand-alone BIM service to cloud BIM service, the cost and concerns associated with BIM adoption completely change. If the organisation is to purchase cloud BIM service from the service providers, the cost for establishing and maintaining local hardware, servers, and dedicated terminals is waived. Instead, users pay monthly subscription fee to acquire the service similar to using the cloud-based services in many other industries such as cloud-based storage service (Dropbox) or the cloud-based video sharing (Vimeo).

Therefore, several cloud-BIM based software and solutions start to come into the construction market. A list of currently available software and service providers to provide the ubiquitous BIM service are provided such as Autodesk BIM 360 Glue, Autodesk BIM 360 Field, Graphisoft BIMcloud, CaddForce, and ITWO 4.0 5D BIM cloud-based construction software. Each of these software supports online collaboration whilst having some key functions that can be utilised by the quantity surveyors. For example, Autodesk BIM 360 Field enables almost anytime, anywhere access to project data throughout the building construction life cycle. It provides standardised checklist templates to track key performance indicators and progress. QS may reply on it during the construction phase for quality control, interim payment management, and filing reports on valuations of variations.

9.3 BIM plus Big data

BIM is a 'richer repository' of information (Eastman, 1999) than a set of drawings or static CAD files. The lack of data and information is frequently claimed by QS as a hindrance for optimum cost estimate. Therefore, with the data accumulated in BIM model and BIM projects, BIM could potential become a hub for big data. Actually, a building information model, regardless of its 3D presentation and information richness, can be perceived as a relational database per se from the perspective of computer database experts (Wang et al., 2013; Liu et al., 2015). Some information is initially contained in the model, whilst others are continuously generated and imported to the model throughout the project life cycle.

BIM has the ability to store different types of information and contains both geometric and non-geometric information about a project (Pratt, 2004). Geometric information includes size, volume, shape, and spatial relationships, whilst non-geometric information includes the type of individual construction component, specifications of material, construction schedule, and cost. Through a certain mapping approach, information from an external source, such as an MS Excel file or MS Access, can be linked to the entire project or a specific component (Xie et al., 2011; Kang and Hong, 2015). With the accumulation of data, the adoption of BIM may be a resource for big data analysis.

It remains controversial whether the BIM-based big data should be protected or shared. The limited access to these data makes data owners to become more protective of their big data and reluctant to share their gold mine with others.

Facebook accumulates big data from its users but only a few individuals have free access to it. Some companies restrict access to their data entirely, others sell access for a fee (Boyd and Crawford, 2012). The open data movement around the world may offset the effects of this trend to a certain extent by calling for big data to be openly available.

Currently, many big data sets are left over unintentionally when businesses are done. For example, they are created as by-products of people travelling around, communicating using smart phones, or purchasing from supermarket or through e-commerce. Likewise, the big data set contained in BIM is a by-product of the BIM adoption process. The amassed data can be a corporate asset, the mining of which allows companies to make better business predictions and decisions. Big data is like a gold mine. Since it is incidentally created and describes natural business processes and captures revealed behaviour, big data tends to be considered better than experimental data or simulation data as it potentially contains more ground truth with respect to social reality than traditional instruments (Hand, 2015). Big data portrays a fuller picture of a subject matter, which allows for a stronger claim to objective truth; as Anderson (2008) put it, "with enough data, the numbers speak for themselves". Industry practitioners are therefore abandoning carefully curated small data and are rushing to discover big data sources to exploit.

9.4 Computational BIM

BIM provides a powerful way to harness the computational power to allow more precise cost planning in the design stage. Computational BIM is the application of computational strategies into the use of BIM. The concept of computational BIM is tightly related to parametric design, algorithmic design, and generative design where the aesthetic elements in design are produced in a computational form and in relation to cost implications.

Computational BIM has several advantages, of which the first and foremost one is that it enables exploring multiple design options with cost implications. By associating the unit cost with BIM elements, change of design can be directly reflected in change of cost. This is particularly useful for design with complexity. For projects with complex visual design such as sculptural forms and intricate façade, a small design change could lead to large variations in QTO and cost planning. The time needed for manual remeasurement and calculation could largely limit the possible design options to be considered. Computational BIM offers a more efficient way to switch between design options, no matter how complex the design is. QS can benefit from the automated measurement and calculation to handle the complexity.

Computational BIM also supports better data management and simulation. By linking the BIM elements with the plug-in of other data management tool such as Excel, the parameters associated with the BIM elements can be exported for further analysis and storage. Through the design process, computational BIM makes it easier to simulate building performance in predicting the operating cost in the building maintenance stage. For example, by conducting the daylight simulation, the daily lighting expense can be estimated. Whilst simulation data is no substitute for actual, real-life data, it does provide a means to evaluate designs based on similar criteria. Quickly determining which design performs measurably better than the others serves as a solid evidence in justifying the choices between design options under the preset cost ceiling.

9.5 New quantity surveyors

The rapid development of BIM will definitely bring impacts and changes to QS with both challenges and opportunities. Referring to the cost-related work and responsibilities of QS at different construction stages (Figure 1.1), BIM could bring a variety of benefits to the work of quantity surveyors at different stages in the new QS business (Figure 9.1).

Preliminary cost estimate: To conduct preliminary cost estimate at the project preparation stages, QS will undertake studies to advise on the feasibility and profitability of the project. Usually, the land acquisition fee, construction cost, maintenance and operating cost, site servicing, cash flows, market analysis, and revenue forecast will be taken into considerations. Previously, quantity surveyors will draw on accumulated experience and recorded cost data to establish a budget based on the projects with similar nature and characteristics. With the historical data stored in existing BIM projects, quantity surveyors could make use of cloud BIM to retrieve the associated data to provide a more accurate estimate.

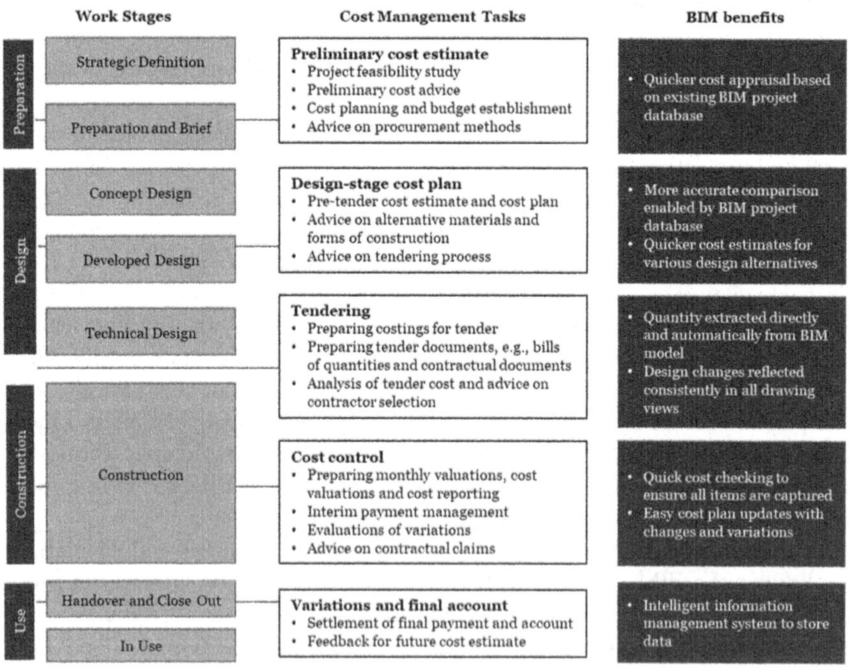

Figure 9.1 Improvements on cost-related work of QS with BIM facilitation.

Design-stage cost plan: The advice of quantity surveyors in the design stage can have profound cost implications for the project. Quantity surveyors need to conduct comparative cost estimate to avoid actual rework in later stages. With the help of BIM, similar projects can be retrieved from the database to enable more accurate cost estimate. Besides, the cost implication of alternative design can also be stored in BIM database, thus allowing quicker recommendations for design alternatives. The design alternatives can also be visualised in BIM, making it easier for quantity surveyors to convey their advice to client and architects.

Tendering: Preparing BoQ is the major responsibility of quantity surveyors in the tendering stage. Quantities of materials, labour work, fees to cover contingency, and any other items that may incur costs during the construction process will be included in the BoQ. The BoQ forms an important part of the contract documents in that it provides a common basis for evaluating and comparing the tenders. With the geometric parameters associated with each component in BIM, the quantities can be extracted directly and automatically from the BIM model (Kymmell, 2008; Jiang, 2011). More importantly, with the automatic quantification function of BIM, quantity surveyors don't need to do manual remeasurement every time when there are design changes. The design changes can be reflected consistently in all drawing views as well as in the calculated quantities.

Cost control: The focus of quantity surveyors' work shifts from cost estimate to cost control during the construction stage. To ensure the expenditure is within the budget limit, quantity surveyors will regularly check upon site construction progress and prepare monthly valuations and cost reporting. The BIM could provide a quick cost checking to ensure all items are captured and process in progress. The quantity surveyors are also the key coordinators between the main contractor, the principal agent, and the client to settle the interim payment. When the cost information is associated with each BIM component, once the BIM model is updated, the cost plans can be easily updated with the associated changes and variations.

Variations and final account: After the construction work is done, quantity surveyors will settle the final payment and final account in the similar way the interim payment is done. To facilitate future cost estimate, the cost data could be stored as a reference. As BIM itself could serve as a database, all the cost information can be stored with the BIM model and BIM projects for future use.

9.6 New QS businesses

The conventional QS profession has been criticised since there is hardly an estimate without its own peculiarities (Sutrisna et al., 2005). The measurement and QTO supported by BIM represents a new approach that could not only speed up the traditional process of BoQ preparation, but also form the basis of a more accurate cost estimate by linking the quantities with the internal built-in or external cost database (Wu et al., 2014). As the role of quantity surveyors will be continuously evolving, BIM is rather enhancing the QS business than eliminating it.

With a series of functions and benefits of BIM to assist quantity surveyors in cost-related tasks, it is also identified that quantity surveyors may encounter difficulties in taking full advantage of BIM due to the technical requirement for BIM operations. Even with the help of BIM, the BoQ cannot be generated with a single click; the quantity surveyor still needs to ensure the calculation rules are aligned with the required standards. When the manual measurement and QTO are waived by the calculation functions in BIM (Wu et al., 2014), these functions in BIM software or platform are still relied on the operations of people. The quantity surveyors may encounter issues such as inconsistent quality of BIM models, inconsistent level of design information included, data exchange issues in BIM tools, and inconsistent formats used for estimating (Wu et al., 2014). Thus, the capability of operating BIM software for QTO and cost management might become the new requirement for surveyors. Meanwhile, whilst BIM could offer help in cost-related task for quantity surveyors, its application in contractual management is still limited. Then quantity surveyors are still indispensable for their competencies in handling contractual issues. The shift of responsibilities enabled by BIM would spur the quantity surveyors to rethink their core competencies.

Meanwhile, there are considerable challenges for the full adoption of BIM in surveying business, amongst which the biggest one is how to filter BIM data to comply with relevant SMM (Olatunji et al., 2010). SMMs fundamentally stipulate the rules and requirement in the preparation of BoQ, which is the basic standard that should not be bypassed no matter what BIM software or BIM solution is used. In addition, the SMMs vary in different regions and different types of construction work. For example, the measurement of building work in Hong Kong is governed by HKSMM4 whilst that in UK is government by New Rules of Measurement volume 2. For the measurement of civil engineering in the UK, however, should comply with the Civil Engineering Standard Method of Measurement volume 3. The variety of SMM requires the rules of QTO and calculation can be comprehensively inclusive to cover all the prevalent SMM, as well as can be flexibly switched when the project locations/types are changed. When the SMM keeps updating, the BIM software and solution providers should also keep pace with it.

9.7 Summary

Computational BIM and cloud BIM represent two growing industry trends in future BIM implementation. Several BIM software and solution providers have developed advantageous functions in computational BIM and cloud BIM to assist QS. With the data accumulated in a BIM model and BIM projects, BIM could be developed into a hub for big data. Data generated and stored in BIM burgeons when multiple parties contribute to the same model, and the dataset demonstrates the ability to account for the totality of the subject under investigation.

Exercising BIM can reshape current quantity surveyors' cost-related tasks, as well as the surveying business at large. However, convenience in quantification and cost estimation comes with some costs. QS will need to take the time to understand and practise the new software, whilst BIM software and solutions providers may have to improve the comprehensiveness and flexibility of their products in order to comply with various forms of SMM.

10 Conclusion

10.1 A sea change in the long-standing QS profession

Over the past 170 years or so, QS as a profession has matured from the UK construction industry and spread to other countries in the world. Whenever there is a construction project, you will see quantity surveyors or construction cost managers, as efficient cost management is of paramount importance to the success or failure of any project. Although changes in the global construction business environment have never ceased, QS has evolved into a profession with a systematic methodology, a stable set of rigorous practices, and a clear definition of its core competencies. Generally, QS plays a key role in relation to cost of a construction project. Without the success of QS, we would not have been possible to develop our built environment, which is instrumental to our human health, economic activity, as well as social behaviour.

Nevertheless, the long-standing profession is also suffering from a series of problems. For example, it requires the techniques to measure quantities from the drawings (knows as QTO), to prepare BoQ, and to check the estimate, which unfortunately form the most time-consuming, tedious, and error-prone part of a QS' works. Surveyors are nowadays often under the time pressure to deliver the tender documents if they serve as the client's/designer's consultant, or to prepare for the bidding if they are on the construction side. They seem having no life, being low-tech, and despised by other professions involved in the AEC industry. The profession has an image problem. This resonated with the RICS (1998), which discovered that many construction clients are critical of traditional QS services and are demanding a different and more comprehensive range of services that is proactive, customer orientated, and supported by significantly better management and business skills. The emerging of new pervasive technologies, such as BIM, big data, and AI provide both the opportunities and challenges to confront the problems.

10.2 BIM technologies demystified

Growing interest in BIM amongst governments, policy-makers, business executives, building professionals, and researchers has rapidly extended BIM's presence in the AEC industry. BIM is soon emerging as a disruptive technology in the global AEC industry. BIM, according to various definitions, is not more than just a digital representation of physical and functional characteristics of a facility. It is the 'digital twin' to map the physical world to form the so-called 'cyber physical system (CPS)' – a concept that is increasingly popular in various scientific realms such as advanced manufacturing, aeronautics and space exploration, and AEC. Many AEC activities (e.g., different design options, construction schemes) can be tried and simulated in the cyber world before they can be applied back to the physical world; a relatively less expensive way. BIM is helping the bidirectional information flow between the cyber and physical worlds.

BIM is materialised in BIM software packages/solutions. Many people mistakenly equate BIM with such software packages/solutions themselves. Many software packages can be called BIM software and be developed to target different professions, such as architecture, structural, MEP engineering, project management, cost management, and FM and could be further extended to other aspects. Although most claim to offer a complete BIM solution, the software packages actually have their own strengths and weaknesses that can be supplemented each other by the companies.

The keyword of BIM is 'information'. Without the 'information', BIM itself can perform very little, if at all. Researchers have proposed different taxonomies of information included in BIM, for example, geometric vs. non-geometric information, or geometric, semantic vs. topological information, or simply all are semantic information. In early stage, information is mainly geometric plus some limited time or cost information, but in recent years, semantic information concerning scopes, requirements, design, production, schedule and plan, maintenance, altering, and demolition is desired to make BIM a truly useful decision-support platform. There are different standards in relation to BIM semantics whilst in a broader sense, LoD is used as a common language to define the information containing in a BIM. The information of BIM is normally stored in various BIM libraries.

At the risk of over simplification, developing a BIM is similar to assembling LEGO. Designers choose different objects and components from the BIM libraries and through many reiterations and finally assemble them into a digital representation of a physical facility to be built. The efficiency of BIM is thus dependent on the designers' capability, the availability of the applicable BIM libraries, and many BIM-related standards (e.g., naming, coding, information enrichment).

10.3 BIM as a disruptor

With that digital presentation, BIM's great advantage is its potential to allow VDC optimising the process of developing a building and the form and performance of that building from occupancy to the end of its lifecycle. Using BIM, persisting problems of the global AEC industry, such as low productivity, poor quality, cost overrun, and excessive material waste, can be possibly alleviated or solved. This book reviewed the main thrusts of BIM as advocated in the literature. BIM allows multiple parties to work simultaneously within one model in facilitating team communication, faster design output and greater adaptability to changing site conditions, and a single repository for which to accumulate and visualise building and operations data, all of which leads to more inclusive and better-quality design. Architects and engineers can draw on embedded geometric, material, and price data to model faster. Construction managers can work off a living design plan, as well as contribute to it. Full stakeholder participation also translates to more correct equipment and logistics and supply chain. BIM can be used to enhance construction productivity, detect design errors and clashes, improve interoperability and communication, and reduce the fragmentation and discontinuity. BIM is even advocated as the disruptive development that it will bring a paradigmatic change to the global AEC industry.

Nevertheless, BIM implementation is not free from issues. There is an anxiety for the industry to find itself a business case for BIM adoption. If a technology initiative such as BIM is to sustain in a competitive business world, it must have a genuine economic foundation. Following the early stage of identifying anecdotal evidence of BIM's benefits, recent years have seen more studies trying to identify empirical evidence of such benefits and this trend will continue. The BIM execution plan is of vital importance to promote BIM implementation in real-life AEC projects. There are other issues such as contractual framework incorporating BIM, IP rights of the model, risks and liability, and organisational supports including BIM-related process and protocols, none of which can have a quick fix to support BIM to have a wider application in the industry.

10.4 BIM for QS

BIM can help QS practice in almost every key stage. In the preparation or design stage of a project, a QS shall prepare a preliminary estimating and cost planning based on design brief, and advice for procurement method suitable for the project. A schematic BIM can be developed to help develop the preliminary cost estimation based on the cost data from database. The schematic BIM will be further developed into the as-designed BIM which contains the consolidated design information. In the tendering stage, a QS shall prepare for BoQ, tender documents, tender reports, and contract documents. BIM can help develop BoQ by automatically extracting the necessary information from the design. BIM can help provide accurate BoQ and other contractual documents for tendering.

In the construction stage, a QS shall provide services like valuation of works for interim, evaluations and certificates, contractual advice, and prepare final account after completion of construction stage. The as-designed BIM will be updated to an as-built BIM by synchronising it with the actual construction information. The as-built BIM can help track construction process and variations, based on which monthly cost reports can be developed and the interim payment can be settled. On completion of the construction work, an as-is or as-built BIM will be developed to reflect the current as-is condition of the constructed project. The as-is BIM provides all necessary information to generate the final account for the settlement of final account. The model can also be used for 'soft landing' in the subsequent FM stage.

The authors of this book developed a QS-BIM execution plan as one of the most significant original contributions. The proposed QS-BIM execution plan consists of four elements to implement BIM for cost management tasks throughout the construction project life cycle, including BIM objectives, procedure, document/information flow, and software/hardware choice and the corresponding responsibilities of the QS. It is not aiming to provide prescribed solutions, but prospective executives of QS firms can use it to tailor a more suitable plan for their firms.

Another important contribution of this book is to demand and define a QS-BIM. A QS-BIM, above all, is a part of the integrated information hub for a construction project but with a focus on the necessary information for quantity surveying tasks. The book explored some of the requirements of a QS-BIM, for example, it should contain the explicit, detailed information of the high-value cost items as well as more detailed information including the architectural, structural, and building services/finishes information. It should comply with the standard rules of standard rules of measurement, specifications, and standards used in a specific local area. Yet, more comprehensive requirements of the QS-BIM are subject to further development.

This book also discussed the critical success factors of BIM adoption for QS, for example, developing the necessary information, compatible with current QS practices, and BIM-based QS solutions. Adoption of BIM for QS should not advocate a radical change of existing, readily accepted practices. Rather, BIM is to serve as a facilitating technology to help QS in a non-intrusive way.

The book further provided three case studies relating to automated QTO, preparing tendering document, and remeasurement of material quantities (e.g., rebars), all are typical QS businesses but with the support of BIM this time. One can perceive them as the implementation of the QS-BIM execution plan in real-life projects. These case studies are not necessarily best practices. Actually, they are far from perfect, but readers may find this part particularly useful as it really delves into BIM-enabled QS, be they successful or not. None of the cases has achieved a holistic BIM. However, they really encourage users to implement BIM in their most promising areas and tackle real problems.

10.5 Big data for QS

Unlike the stereotype that the AEC industry is 'small' in data generation caused by the widespread fragmentation and discontinuity in project organisations, it is discovered that the industry is actually 'rich' in data. Numerous data, with the characteristics of volume, velocity, and variety, are continuously generated from different AEC activities throughout a project life cycle. However, the big data is just discarded, or it is just residing in different companies, whose power is yet to be fully exploited.

This book provides three cases in which big data can make a non-trivial contribution to construction cost management. For example, big data can help clients and their cost consultants to prepare tendering and cost estimate in a faster and more accurate way. Big data can help contractors and subcontractors to prepare their bidding and pricing more efficiently and accurately. Big data can also help clients and their cost consultants to analyse bidders' behaviour, which is a significant issue in today's competitive bidding and tendering practices. Other than the three cases, big data can be used in other QS business, such as cost control, variations, and final account. For example, big data can help recognise the critical points for variations; therefore, special cost control attentions can be paid to the points. Properly harnessed, big data can allow cost consultants to give a reasonably accurate estimate to their clients even when the design details are missing or unavailable.

However, big data is not a silver bullet to cure all the issues in QS. Big data, in comparison with 'small' data, is just able to provide a fuller picture of a subject to allow a closer access to the totality of the subject. If we treat QS as making an array of decisions, with the access to the totality, big data can help alleviate the potential bias in making these decisions. Big data analytics can also help identify some hidden patterns, unknown correlations, and other useful information (e.g., the bidders' behaviour) in making more informed decisions in QS business. In this book, we took Simon's (1982) bounded rationality as the theoretical underpinning to explain the vogue of big data. Big data can help increase our rationality which is bounded by limited information, cognitive capability, and time we have to make a decision.

AEC big data existed before the popularisation of BIM, but the introduction of BIM as the 'disruptor' of the industry provided an unprecedented opportunity to have big data and unleash its power. BIM models are important components of big data for construction cost management, which can be further integrated with a large variety of project information including the geometric, semantic, and topological information updated promptly. BIM also became a shared platform based on which big data can be accumulated, contained, and exploited to form a more reliable "Single Source of Truth" to support stakeholders' decision-making throughout AEC stages, including construction cost management. In a sense, big data and BIM help alleviate the fundamental problems of fragmentation and discontinuity in this industry.

10.6 Challenges of BIM and big data for QS

Despite the impressive momentum that has been achieved in implementing BIM and big data for QS, there are still many challenges ahead before they can be a natural ally of QS. This book discussed several categories of challenges that are relating to standards, technology, economic, organisational issues, and the alike. Construction is an industry particularly relying on various well-defined standards to regulate its activities, procedures, and final products. Owing to the great efforts, a succession of BIM-related standards has been published in major economies, which has largely promoted BIM implementation over the past decade. Nevertheless, it is yet to reach the extent to which BIM standards are readily operable for construction cost management. More efforts are desired to further detail the standards in terms of file format, naming convention, coding, information, LoD, information exchange, contractual relationships, liability, amongst the many other things.

Technical challenges are yet to be fully overcome although much has been achieved in BIM technologies over the past decade. Different companies have their own preferred BIM software solutions, and their interoperability seems being undermined rather than improved over the past decade. Modelling technologies are still relatively burdensome; BIM users still need to take extra, sometimes excessive efforts to develop a QS-BIM; therefore, they are unwilling to do it. Hardware still finds itself 'sticky' in manipulating BIM with LoD 400 or above. There is no certain answer whether the cost and quantity data should be embedded with BIM objects, or placed in a gateway. BIM is increasingly migrating to cloud services, but current network bandwidth is often a bottleneck. To have a 4D to nD BIM is critical, but the technologies (e.g., laser scanning, photogrammetry) to capture real-time site conditions and form the as-built BIM for cost management are yet to mature. How to manage BIM and big data is a challenge that has not been seen in the small data era. Cybersecurity is also a technical challenge in BIM progression.

Economic challenges are still holding when it comes to BIM and big data for QS. The cost comprises of the capital investment in hardware, software, and training, and the ongoing investment for updating them, without mentioning other opportunity cost and risks. The cost could be overwhelming to SMEs, which often have limited resources in implementing BIM. Although many academic studies have reported that the cost of BIM implementation can be offset by the benefits it brings, its economic legitimacy is yet to be widely seen in real-life AEC projects, and backed up by more empirical and anecdotal evidence. More research efforts are desired to analyse the cost and benefits associating with BIM and big data for QS. The difficulties to conduct such CBA are also apparent. It is not easy to detach the part of benefits that is contributed by BIM, particularly when BIM is closely embedded in business process. The data for a proper CBA are not readily accessible. Furthermore, the authors of this book advise a proper cost and benefit redistribution amongst the parties, for example, the client, the designer, and the surveyor, so that all the participants are incentivised to consummately implement BIM and harness its full power. Current fee structure may not have reflected this need.

No BIM is operating in a vacuum; the success of BIM cannot be disconnected with its organisational structure, which could be at the company level as a long-standing organisation, or at the project level as a temporary delivery organisation. Implementing BIM will unavoidably change reporting structures, communication patterns, and regular working processes; therefore, the resistance to change is considerably overwhelming. BIM is a sensible way to alleviate the fragmentation and discontinuity caused by the traditional DBB, which is still a prevailing procurement model. Nevertheless, it is somewhat unrealistic to hope that BIM, a technology, can change the structural problem in organisations. BIM should be working with an amenable procurement model (e.g., IPD) to unleash its full power in construction cost management.

Legal and liability issues associated with BIM need to be sorted out for QS business. The use of BIM encourages the seamless collaboration between all stakeholders, which, however, may blur the level of responsibility amongst individual stakeholders and the assignment of liability issues amongst them. Several legal issues should be considered, which include but not limited to the ownership of the BIM model, the use and distribution of the BIM model, and its IP rights, its liability when things unfortunately go wrong, and how BIM can be properly bound in contractual requirements. A new form of contract could be implemented in order to avoid arguments concerning BIM responsibilities and liabilities. Yet, this contract is not fully developed and accepted by the industry.

Cultural challenges affect the implementation of BIM and big data for construction cost management. The industry is notorious for being slow in embracing changes and new technologies. To be fair, it is probably prudential for the industry to stick to the traditional technologies and methods to deal with the heterogeneity of AEC projects, time pressure to deliver, uncertainty of the market, and risks to adopt disruptive approaches. Nevertheless, cultural hindrances such as mindset and prevailing conservatism provide an explanation to the sluggish intake of BIM in QS business. In parallel with cultural challenges is a lack of knowledge in the industry. There is a deficiency of BIM and big data expertise when it comes to their applications for cost management. All these accumulated together to explain why QS favour the conventional methods that have been utilised for decades and are reluctant to change to the BIM and big data era.

10.7 Recommendable good practices

In face of the challenges, readers and prospective BIM users, however, should not be overwhelmed or feel pessimistic about the future of BIM and big data for QS. This book recommends the good practices that can be used to address the above challenges. Seven categorises of good practices are recommended, including (i) Encouraging R&D, (ii) Training and education, (iii) Making a strong business case, (iv) Cost-benefit sharing, (v) Embracing innovative procurement models, (vi) Localising BIM, and (vii) 'Think big, act small'. However, it should be bored in mind that there is no perfect solution or quick fix to these challenges. These good practices are achievable in long rather than short term.

Continuous R&D is an important strategy for companies to sharpen their competitive advantages by developing new services or improving existing services to the clients. BIM and big data technologies are evolving rapidly. R&D is essential to understand the technologies and develop their related standards, code of practice, and the like. R&D is indispensable to continuously improve the knowledge and AI competencies of the profession to provide improved QS and cost management services. With big data and BIM accumulated, companies are able to provide better VFM in QS services. The industry is now investing in cutting-edge technologies such as VR/AR and 3D scanning for BIM and QS. R&D has also been focusing on AI such as ML and DL to mine the big data of BIM and QS.

BIM and big data like any emerging technology come with a learning curve. Construction industry practitioners require training and practical experience before they can operate BIM software with dexterity. Training and education ensure the implementation of state-of-the-art technologies in QS. Knowledge sharing programmes should also be added to the training system. Encouragingly, these training programmes have been seen in the CPD carried out by various professional bodies. They also work with universities to provide education to the young generation to equip them with BIM and big data knowledge before they join the AEC industry.

BIM and big data for QS business need to have a genuine economic legitimacy to be sustained in a business world, as they often compete against other initiatives for companies' scarce financial resources. To address the economic challenges facing BIM and big data, it is critical to make a strong business case. Companies should have a deep understanding of the benefits and costs of BIM and big data. They should devise proper strategies, for example, selecting proper hardware and software, reengineering the business process, mobilising discretionary management, and, equally importantly, monitoring the effects to allow that the costs are offset by the benefits in many BIM-enabled QS practices.

In addition, this book advocates further exploration of cost-benefit sharing models, with a view to sustaining BIM in QS business. Costs and benefits mean different things to different project stakeholders, which are largely different profit centres inherently safeguarding their own interests. Whatever costs and benefits materialised in a project as a collaborative environment will be ultimately channelled back to the participants, ideally in a way that can incentivised the project stakeholders to embrace BIM consummately. Currently, this is largely reflected in a fee structure, for example, to specify how much a profession can normally take from a project's overall investment. Prevailing fee structures are largely based on old patterns without BIM and big data. They should be revisited in the BIM and big data era.

To address the organisational issues that may constrain the adoption of BIM and big data for QS, the book recommends embracing innovative procurement models such as D&B, New Engineering Contract (NEC), IPD, or more collaborative models to be invented. As IPD and BIM share similar principles of collaboration, the fusion of BIM and IPD is gradually changing the working patterns of building projects. Yet, the industry is encouraged to continuously innovate its procurement models that can better deliver the value of construction.

In view of the challenges of lack of operable standards, relative high cost in developing BIM, and so on, this book recommends 'BIM localisation'. It is noticed that current BIM technologies, organisational settings, and related working protocols are developed from a 'standard' background and care little about a local context. This largely explains the challenges as just exhibited and BIM localisation is a process to analyse, adapt, and adopt general BIM knowledge (i.e., BIM technology, process, and policy) to suit the specific needs (e.g., regulations, building codes, and cultures) of a local context.

At last, this book suggests that QS businesses do not have to adopt a comprehensive and ambitious BIM and big data strategy from day one. Rather, they can implement the technologies to solve their most urgent and promising issues, such as automated QTO, preparing for tendering, or forming the cost control baseline. It is recommendable to implement BIM in a smaller scale, and in most promising areas and gradually scale up to the whole company, and to other AEC business after accumulating sufficient experiences.

10.8 A bright future of BIM and big data for QS

After a thorough discussion, it is becoming clear that BIM and big data have a bright future for construction cost management despite many challenges and hurdles exist. BIM has moved quite a long way from traditional 'clash detection' to 'optimisation', QTO, BoQ generation, and now 5D BIM for better cost management with the big data and experiences accumulated. A clear trend is that BIM is migrating to cloud and becoming ubiquitously available for BIM services, including QS/cost management. AI enhanced by big data is increasingly explored to replace human QS in some tedious, error-prone, and repetitive works so QS can be freed to become consultants to advice on macro or more strategic cost issues. To this end, BIM and big data are not replacing QS. Rather, this long-standing profession is facing unprecedented opportunities to reinvent its core competencies, to elevate its professional image, and to contribute more values than the profession has honourably contributed over the past century.

References

AEC (UK). (2015). *BIM Technology Protocol.* Available at: https://aecuk.wordpress.com/documents/

Aibinu, A., and Venkatesh, S. (2013). Status of BIM adoption and the BIM experience of cost consultants in Australia. *Journal of Professional Issues in Engineering Education and Practice, 140*(3), 04013021.

Agrawal, A. (2006). Engaging the inventor: Exploring licensing strategies for university inventions and the role of latent knowledge. *Strategic Management Journal, 27*(1), 63–79.

Akinsiku, E. O., Babatunde, S. O., and Opawole, A. (2011). Comparative accuracy of floor area, storey enclosure and cubic methods in preparing preliminary estimate in Nigeria. *Journal of Building Appraisal*, 6(3–4), 315–322.

Alavi, M., and Leidner, D. E. (2001). Knowledge management and knowledge management systems: Conceptual foundations and research issues. *MIS Quarterly, 25*(1), 107–136.

Alazmeh, N. I. D. A. A., Underwood, J. A. S. O. N., and Coates, S. P. (2017). Implementing a BIM collaborative workflow in the UK construction market. *International Journal of Sustainable Development and Planning, 13*(1), 24–35.

Amendola, A. (2002). Recent paradigms for risk informed decision making. *Safety Science, 40*(1–4), 17–30.

American Institute of Architects (AIA). (2007a). *Integrated Project Delivery: A Guide.* Available at: https://info.aia.org/SiteObjects/files/IPD_Guide_2007.pdf

American Institute of Architects (AIA). (2007b). *A Working Definition: Integrated Project Delivery.* Available at: http://aiacc.org/wp-content/uploads/2010/07/A-Working-Definition-V2-final.pdf

American Institute of Architects (AIA). (2013a). *G202–2013 Project Building Information Modelling Protocol Form.* Available at: https://www.aiacontracts.org/contract-documents/19016-project-bim-protocol

American Institute of Architects (AIA). (2013b). *E203–2013 Building Information Modeling and Digital Data Exhibit.* Available at: https://www.aiacontracts.org/contract-documents/19026-building-information-modeling-and-digital-data-exhibit

Anderson, C. (2008). *The End of Theory: The Data Deluge Makes the Scientific Method Obsolete. WIRED.* Available at: https://www.wired.com/2008/06/pb-theory/

Arayici, Y., Coates, P., Koskela, L., Kagioglou, M., Usher, C., and O'reilly, K. (2011). Technology adoption in the BIM implementation for lean architectural practice. *Automation in Construction, 20*(2), 189–195.

Arensman, D. B., and Ozbek, M. E. (2012). Building information modelling and potential legal issues. *International Journal of Construction Education and Research*, 8(2), 146–156.

Ashworth, A. (2010). *Cost Studies of Buildings* (5th ed.). Harlow: Prentice Hall.

Ashworth, A., and Hogg, K. (2007). *Willis's Practice and Procedure for the Quantity Surveyor* (12th ed.). Oxford: Blackwell Publishing Ltd.

Association of Construction and Development (ACD). (2012). *Clash Detection in BIM Modelling.* Available at: http://www.associationofconstructionanddevelopment.org/articles/view.php?article_id=10780

Autodesk. (2007). *BIM and Cost Estimating.* Available at: http://www.consortech.com/bim2/documents/BIM_cost_estimating_EN.pdf

Azhar, S. (2011). Building information modelling (BIM): Trends, benefits, risks, and challenges for the AEC industry. *Leadership and Management in Engineering, 11*(3), 241–252.

Barlish, K., and Sullivan, K. (2012). How to measure the benefits of BIM—A case study approach. *Automation in Construction, 24,* 149–159.

Becerik-Gerber, B., and Rice, S. (2010). The perceived value of building information modelling in the US building industry. *Journal of Information Technology in Construction, 15*(2), 185–201.

Bekker, H., Thornton, J. G., Airey, C. M., Connelly, J. B., Hewison, J., Robinson, M. B., and Pearman, A. D. (1999). Informed decision making: An annotated bibliography and systematic review. *Health Technology Assessment, 3*(1), 1–156.

Bernstein, H. M. (2003). Measuring productivity: An industry challenge. *Civil Engineering, 73*(12), 46–53.

BIMForum. (2017). *Level of Development Specification Part 1.* Available at: http://bimforum.org/wp-content/uploads/2017/11/LOD-Spec-2017-Part-I-2017-11-07-1.pdf

Boyd, D., and Crawford, K. (2012). Critical questions for big data: Provocations for a cultural, technological, and scholarly phenomenon. *Information, Communication & Society, 15*(5), 662–679.

Bresnen, M., Goussevskaia, A., and Swan, J. (2004). Embedding new management knowledge in project-based organizations. *Organization Studies, 25*(9), 1535–1555.

Bryde, D., Broquetas, M., and Volm, J. M. (2013). The project benefits of building information modelling (BIM). *International Journal of Project Management, 31*(7), 971–980.

buildingSMART. (2016). *IFC4 Addendum 2.* Available at: http://www.buildingsmart-tech.org/specifications/ifc-releases/ifc4-add2

Carlsson, S. A., El Sawy, O., Eriksson, I., and Raven, A. (1996). Gaining competitive advantage through shared knowledge creation: In search of a new design theory for strategic information systems. In *Proceedings of the Fourth European Conference on Information Systems,* July 2–4, 1996 in Lisbon, Portugal (pp. 1067–1076).

Cartlidge, D. (2009). *Quantity Surveyor's Pocket Book.* Abingdon: Routledge.

Cartlidge, D. (2011). *New Aspects of Quantity Surveying Practice.* Oxon: Spon Press.

Chang, C. Y., and Howard, R. (2014). An economic framework for analyzing the incentive problems in building information modeling systems. In *Proceedings of Academy of Management Annual Meeting,* Philadelphia, August 9–13, 2013. New York: Academy of Management.

Chen, D., Doumeingts, G., and Vernadat, F. (2008). Architectures for enterprise integration and interoperability: Past, present and future. *Computers in Industry, 59*(7), 647–659.

Chen, K., Lu, W., Peng, Y., Rowlinson, S., and Huang, G. Q. (2015). Bridging BIM and building: From a literature review to an integrated conceptual framework. *International Journal of Project Management, 33*(6), 1405–1416.

Chen, K., Lu, W., Wang, H., Niu, Y., and Huang, G. G. (2017). Naming objects in BIM: A convention and a semiautomatic approach. *Journal of Construction Engineering and Management, 143*(7), 06017001.

Chen, J., Chen, Y., Du, X., Li, C., Lu, J., Zhao, S., and Zhou, X. (2013). Big data challenge: A data management perspective. *Frontiers of Computer Science*, *7*(2), 157–164.

Cheng, J. C., and Lu, Q. (2015). A review of the efforts and roles of the public sector for BIM adoption worldwide. *Journal of Information Technology in Construction*, *20*(27), 442–478.

Cheng, T., and Teizer, J. (2013). Real-time resource location data collection and visualization technology for construction safety and activity monitoring applications. *Automation in Construction*, *34*, 3–15.

Chien, K. F., Wu, Z. H., and Huang, S. C. (2014). Identifying and assessing critical risk factors for BIM projects: Empirical study. *Automation in Construction*, *45*, 1–15.

Chien, S. C., Chuang, T. C., Yu, H. S., Han, Y., Soong, B. H., and Tseng, K. J. (2017). Implementation of cloud BIM-based platform towards high-performance building services. *Procedia Environmental Sciences*, *38*, 436–444.

Cidik, M. S., Boyd, D., and Thurairajah, N. (2014). Leveraging collaboration through the use of building information models. In *Proceedings 30th Annual Association of Researchers in Construction Management Conference*, Portsmouth, September 1–3. UK: Association of Researchers in Construction Management (pp. 713–722).

Clark, J. H. (1976). Hierarchical geometric models for visible surface algorithms. *Communications of the ACM*, *19*(10), 547–554.

Coates, L. C., Fransen, J., and Helliwell, P. S. (2010). Defining minimal disease activity in psoriatic arthritis: A proposed objective target for treatment. *Annals of the Rheumatic Diseases*, 69(01), 48–53.

Computer Integrated Construction Research Group (CICRG). (2010). *BIM Project Execution Planning Guide—Version 2.0*. The Pennsylvania State University, University Park, PA, USA.

Construction Industry Council of United Kingdom (UKCIC). (2013). *Building Information Model (BIM) Protocol: Standard Protocol for Use in Projects Using Building Information Models* (2nd ed.). Available at: http://cic.org.uk/admin/resources/bim-protocol2nd-edition-1.pdf

Construction Specifications Institute (CSI). (2011). *Construction Specifications Practice Guide*. Hoboken, NJ: Wiley.

Dainty, A., Murray, M., and Moore, D. (2007). *Communication in Construction: Theory and Practice*. Abingdon: Routledge.

Davies, R., and Harty, C. (2011). Building Information Modelling as innovation journey: BIM experiences on a major UK healthcare infrastructure project. In *6th Nordic Conference on Construction Economics and Organisation—Shaping the Construction/Society Nexus*, Copenhagen, April 13–15, 2011. Hørsholm: SBI forlag.

Davtalab, O., and Delgado, J. L. (2014) Benefits of 6D BIM for facilities management departments for construction projects—A case study approach. In *Proceedings of the International Symposium on Automation and Robotics in Construction and Mining*, Sydney, July 9–11.

DPR Construction. (2010). *A Case Study in Work Environment Redesign*. Available at: http://dupress.deloitte.com/dup-us-en/topics/talent/dpr-construction.html

Dretske, F. (1981). *Knowledge and the Flow of Information*. Cambridge, MA: MIT Press.

Eastman, C. M. (1999). *Building Product Models: Computer Environments, Supporting Design and Construction*. London: CRC Press.

Eastman, C. M., Jeong, Y. S., Sacks, R., and Kaner, I. (2009). Exchange model and exchange object concepts for implementation of national BIM standards. *Journal of Computing in Civil Engineering*, *24*(1), 25–34.

Eastman, C. M., Teicholz, P., Sacks, R., and Liston, K. (2011). *BIM Handbook: A Guide to Building Information Modelling for Owners, Managers, Designers, Engineers and Contractors*. Hoboken, NJ: John Wiley & Sons.

Egan, J. (1998). *Rethinking Construction. The Report of the Construction Task Force*. Available at: http://constructingexcellence.org.uk/wp-content/uploads/2014/10/rethinking_construction_report.pdf
Elmualim, A., and Gilder, J. (2014). BIM: Innovation in design management, influence and challenges of implementation. *Architectural Engineering and Design Management*, *10*(3–4), 183–199.
El-Omari, S., and Moselhi, O. (2011). Integrating automated data acquisition technologies for progress reporting of construction projects. *Automation in Construction*, *20*(6), 699–705.
Escamilla, E., Ostadalimakhmalbaf, M., and Bigelow, B. F. (2016). Factors impacting Hispanic high school students and how to best reach them for the careers in the construction industry. *International Journal of Construction Education and Research*, *12*(2), 82–98.
Flannery, R. (2015). *Glodon Software Entrepreneur joins China's billionaire ranks*. Available at: http://www.forbes.com/sites/russellflannery/2015/04/03/glodon-software-entrepreneur-joins-chinas-billionaire-ranks/#3fd897fc499d
Flanagan, R., and Lu, W. (2008) Making informed decisions in product-service systems. In *IMechE Conference, Knowledge and Information Management Through-Life*. Institute of Mechanical Engineers, London.
Forgues, D., and Iordanova, I. (2010). An IDP-BIM framework for reshaping professional design practices. In *Proceedings of Construction Research Congress 2010: Innovation for Reshaping Construction Practice*, Alberta, May 8–10, 2010. Teston, VA: American Society of Civil Engineers (pp. 172–182).
Forgues, D., Iordanova, I., Valdivesio, F., and Staub-French, S. (2012). Rethinking the cost estimating process through 5D BIM: A case study. In *Proceedings of Construction Research Congress 2012*, West Lafayette, May 21–23, 2012. Teston, VA: American Society of Civil Engineers (pp. 778–786).
Fox, S., and Hietanen, J. (2007). Interorganisational use of building information models: Potential for automational, informational and transformational effects. *Construction Management and Economics*, *25*(3), 289–296.
Gallaher, M. P., O'Connor, A. C., Dettbarn, J. L., and Gilday, L. T. (2004). *Cost Analysis of Inadequate Interoperability in the US Capital Facilities Industry*. NIST Publication GCR 04–867. Scotts Valley, CA: CreateSpace Independent Publishing Platform.
Gann, D. M., and Salter, A. J. (2000). Innovation in project-based, service-enhanced firms: The construction of complex products and systems. *Research Policy*, *29*(7–8), 955–972.
General Services Administration (GSA). (2007). *BIM Guide Series*. Available at: https://www.gsa.gov/portal/category/101070
Ghaffarianhoseini, A., Tookey, J., Ghaffarianhoseini, A., Naismith, N., Azhar, S., Efimova, O., and Raahemifar, K. (2017). Building Information Modelling (BIM) uptake: Clear benefits, understanding its implementation, risks and challenges. *Renewable and Sustainable Energy Reviews*, *75*, 1046–1053.
Giel, B. K., and Issa, R. R. (2011). Return on investment analysis of using building information modelling in construction. *Journal of Computing in Civil Engineering*, *27*(5), 511–521.
Goucher, D., and Thurairajah, N. (2013). Advantages and challenges of using BIM: A cost consultant's perspective. Paper presented in *49th ASC Annual International Conference*, San Luis Obispo, CA, April 3–13, 2013.
Gobble, M. M. (2014). Design thinking. *Research-Technology Management*, *57*(3), 59–62.
Grant, R. M. (1996). Toward a knowledge-based theory of the firm. *Strategic Management Journal*, *17*(S2), 109–122.

Gu, N., and London, K. (2010). Understanding and facilitating BIM adoption in the AEC industry. *Automation in Construction, 19*(8), 988–999.

Hampton, S. E., Strasser, C. A., Tewksbury, J. J., Gram, W. K., Budden, A. E., Batcheller, A. L., and Porter, J. H. (2013). Big data and the future of ecology. *Frontiers in Ecology and the Environment, 11*(3), 156–162.

Hand, D. J. (2015). Official statistics in the new data ecosystem. Paper presented in *The New Techniques and Technologies in Statistics Conference*, Brussels, March 10–12, 2015.

Hannon, J. J. (2007). Estimators' functional role change with BIM. AACE *International Transactions*, IT31. Morgantown, WV: AACE International.

Hao, J., Zhu, J., and Zhong, R. (2015). The rise of big data on urban studies and planning practices in China: Review and open research issues. *Journal of Urban Management*, 4(2), 92–124.

Hardin, B., and McCool, D. (2015). *BIM and Construction Management: Proven Tools, Methods, and Workflows*. Hoboken, NJ: John Wiley & Sons.

Harty, C., Throssell, D., Jeffrey, H., and Stagg, M. (2010). Implementing building information modelling: A case study of the Barts and the London hospitals. In *Proceedings of the International Conference on Computing in Civil and Building Engineering*, Nottingham, June 30–July 2, 2010. Nottingham: Nottingham University Press.

Holzer, D. (2015). *The BIM Manager's Handbook, Part 3: Focus on Technology*. Hoboken, NJ: John Wiley & Sons.

Hong Kong Construction Industry Council (HKCIC). (2014). *Roadmap for Building Information Modelling Strategic Implementation in Hong Kong's Construction Industry*. Available at: goo.gl/94MV4J

Hong Kong Construction Industry Council (HKCIC). (2015a). *Project Clients Summit: Development of BIM Implementation Strategies Summit Report for Project Clients from Government Sector*. Available at: http://www.cic.hk/cic_data/pdf/about_cic/publications/eng/reference_materials/HKCIC_FinalReport_GovSector_201511.pdf

Hong Kong Construction Industry Council (HKCIC). (2015b). *Project Clients Summit: Development of BIM Implementation Strategies Summit Report for Project Clients from Private Sector*. Available at: http://www.cic.hk/cic_data/pdf/about_cic/publications/eng/reference_materials/HKCIC_FinalReport_PrivateSector_201511.pdf

Hong Kong Institute of Surveyors (HKIS). (2012a). *Practice Notes for Quantity Surveyors – Cost Control and Financial Statements*. Available at: https://www.hkis.org.hk/ufiles/QS-CostControl2015.pdf

Hong Kong Institute of Surveyors (HKIS). (2012b). *Practice Notes for Quantity Surveyors – Final Accounts*. Available at: https://www.hkis.org.hk/ufiles/QS-Final_Accounts2015.pdf

Hong Kong Institute of Surveyors (HKIS). (2012c). *Practice Notes for Quantity Surveyors – Tendering*. Available at: https://www.hkis.org.hk/hkis/papers/qsd/qsd-pn04.pdf

Hong Kong Institute of Surveyors (HKIS). (2016). *Practice Notes for Quantity Surveyors – Pre-contract Estimates and Cost Plans*. Available at: https://www.hkis.org.hk/ufiles/QS-costplans2016.pdf

Hong Kong Institute of Surveyors (HKIS). (2017). *HKIS Valuation Standards 2017*. Available at: https://www.hkis.org.hk/en/publication_sales.php

Howe, D., Costanzo, M., Fey, P., Gojobori, T., Hannick, L., Hide, W., Hill, D. P., Kania, R., Schaeffer, M., Pierre, S., Twigger, S., White, O., and Rhee, S. Y. (2008). Big data: The future of bio-curation. *Nature, 455* (7209), 47–50.

Hurtado, K., and O'Connor, P. (2008). Contractual issues in the use of building information modelling. *International Construction Law Review, 25*(3), 262–272.

International Organisation for Standardisation (ISO). (2016). *ISO 29481-1:2016 Building Information Models – Information Delivery Manual – Part 1: Methodology and Format*. Available at: http://www.iso.org/iso/catalogue_detail.htm?csnumber=60553

Isikdag, U., Aouad, G., Underwood, J., and Wu, S. (2007). Building information models: A review on storage and exchange mechanisms. Paper presented in the *24th CIB* W78 Conference "Bringing ITC knowledge to work", June 27–29, Maribor, Slovenia.

Jefferies, M. C., and Rowlinson, S. (2016). *New Forms of Procurement: PPP and Relational Contracting in the 21st Century*. Abingdon: Routledge.

Jiang, X. (2011). *Developments in Cost Estimating and Scheduling in BIM Technology*. Boston, MA: Northeastern University.

Jrade, A., and Alkass, S. (2007). Computer-integrated system for estimating the costs of building projects. *Journal of Architectural Engineering, 13*(4), 205–223.

Jung, Y., and Joo, M. (2011). Building information modelling (BIM) framework for practical implementation. *Automation in Construction, 20*(2), 126–133.

Kaner, I., Sacks, R., Kassian, W., and Quitt, T. (2008). Case studies of BIM adoption for precast concrete design by mid-sized structural engineering firms. *Journal of Information Technology in Construction, 13*, 303–323.

Kang, L. S., Moon, H. S., Ji, S. B., and Lee, T. S. (2008). Development of major functions of visualization system for construction schedule data in plant project. *Korea Journal of Construction Engineering and Management*, 9(1), 66–76.

Kang, T. W., and Hong, C. H. (2015). A study on software architecture for effective BIM/GIS-based facilities management data integration. *Automation in Construction, 54*, 25–38.

Kehily, D., and Underwood, J. (2017). Embedding life cycle costing in 5D BIM. *Journal of Information Technology in Construction, 22*, 145–167.

Kim, K. P., and Park, K. S. (2016). *Implication of Quantity Surveying Practice in a BIM-Enabled Environment*. Selangor: The Pacific Association of Quantity Surveyors.

Kim, K., and Teizer, J. (2014). Automatic design and planning of scaffolding systems using building information modelling. *Advanced Engineering Informatics, 28*(1), 66–80.

Kunz, J., and Fischer, M. (2009). Virtual design and construction themes, case studies and implementation suggestions. *CIFE Technical Reports*. Available at: http://www.stanford.edu/group/CIFE/online.publications/WP097.pdf.

Kymmell, W. (2008). *Building Information Modelling: Planning and Managing Construction Projects with 4D CAD and Simulations*. New York: McGraw Hill Professional.

Larson, D., and Golden, K. (2008). Entering the brave new world: An introduction to contracting for BIM. *William Mitchell Law Review, 34*(1), 75–108.

Latham, M. (1994). *Constructing the team: Final report of the government/industry review of procurement and contractual arrangements in the UK construction industry*. London: HMSO.

Lee, S. K., Kim, K. R., and Yu, J. H. (2014). BIM and ontology-based approach for building cost estimation. *Automation in Construction, 41*, 96–105.

Li, H., Lu, W., and Huang, T. (2009). Rethinking project management and exploring virtual design and construction as a potential solution. *Construction Management and Economics, 27*(4), 363–371.

Liang, C., Lu, W., Rowlinson, S., and Zhang, X. (2016). Development of a multifunctional BIM maturity model. *Journal of Construction Engineering and Management, 142*(11), 06016003.

Liu, H., Al-Hussein, M., and Lu, M. (2015). BIM-based integrated approach for detailed construction scheduling under resource constraints. *Automation in Construction, 53*, 29–43.

Lu, N., and Korman, T. (2010). Implementation of building information modelling (BIM) in modular construction: Benefits and challenges. In *Construction Research Congress 2010*, Alberta, May 8–10, 2010. Teston, VA: American Society of Civil Engineers (pp. 1136–1145).

Lu, W., Peng, Y., Shen, Q., and Li, H. (2012). Generic model for measuring benefits of BIM as a learning tool in construction tasks. *Journal of Construction Engineering and Management*, *139*(2), 195–203.

Lu, Q., Wang, J., and Cheng, J. C. P. (2016). A financial decision making framework for construction projects based on 5D Building Information Modeling (BIM). *International Journal of Project Management*, *34*(1), 3–21.

Lu, W., Fung, A., Peng, Y., Liang, C., and Rowlinson, S. (2014). Cost–benefit analysis of Building Information Modelling implementation in building projects through demystification of time-effort distribution curves. *Building and Environment*, *82*, 317–327.

Lu, W., Huang, G. Q., and Li, H. (2011). Scenarios for applying RFID technology in construction project management. *Automation in Construction*, *20*(2), 101–106.

Lu, W., and Li, H. (2011). Building information modelling and changing construction practices. *Automation in Construction*, *20*(2), 99–100.

Lu, W., Peng, Y., Shen, Q., and Li, H. (2012). Generic model for measuring benefits of BIM as a learning tool in construction tasks. *Journal of Construction Engineering and Management*, *139*(2), 195–203.

Lu, W., and Olofsson, T. (2014). Building information modelling and discrete event simulation: Towards an integrated framework. *Automation in Construction*, *44*, 73–83.

Lu, W., Webster, C., Chen, K., Zhang, X., and Chen, X. (2017). Computational BIM for construction waste management: Moving from rhetoric to reality. *Renewable and Sustainable Energy Reviews*, 68(1), 587–595.

Machlup, F. (1983). *The Study of Information: Interdisciplinary Messages*. Wiley.

MacLeamy, P. (2004) *Collaboration, Integrated Information and the Project Lifecycle in Building Design, Construction and Operation*. Available at: http://codebim.com/wp-content/uploads/2013/06/CurtCollaboration.pdf

MacLeamy, P. (2008), *BIM, BAM, BOOM! How to Build Greener, High-Performance Buildings*. Urban Land Green Magazine.

Mahamadu, A. M., Mahdjoubi, L., and Booth, C. (2013). Challenges to BIM-cloud integration: Implication of security issues on secure collaboration. In *5th International Conference on Cloud Computing Technology and Science* (Vol. 2, pp. 209–214). IEEE.

Malleson, A. (2018). *National BIM Survey: Summary of Findings*. Available at: https://www.thenbs.com/knowledge/the-national-bim-report-2018

Masterman, J. (2003). *An Introduction to Building Procurement Systems*. Abingdon: Routledge.

Matipa, W. M., Cunningham, P., and Naik, B. (2010). Assessing the impact of new rules of cost planning on Building Information Modelling (BIM) schema pertinent to quantity surveying practice. In *26th Annual ARCOM Conference*, Leeds, UK.

Matipa, W. M., Kelliher, D., and Keane, M. (2008). How a quantity surveyor can ease cost management at the design stage using a building product model. *Construction Innovation*, 8(3), 164–181.

Mayer-Schönberger, V., and Cukier, K. (2013). *Big Data: A Revolution That Will Transform How We Live, Work, and Think*. Boston, MA: Houghton Mifflin Harcourt.

McAfee, A., Brynjolfsson, E., Davenport, T. H., Patil, D. J., and Barton, D. (2012). Big data: The management revolution. *Harvard Business Review*, 90(10), 60–68.

McAuley, B., Hore, A., and West, R. (2016) BICP Irish BIM Study. *Irish Building Magazine*, *4*, 78–81.
McGraw-Hill Construction. (2007). *Interoperability in the Construction Industry*. Available at: http://www.1stpricing.com/pdf/MGH_Interoperability_SmartMarket_Report.pdf
McGraw-Hill Construction. (2010). *The Business Value of BIM in Europe*. Available at: http://images.autodesk.com/adsk/files/business_value_of_bim_in_europe_smr_final.pdf
McGraw-Hill Construction. (2012). *The Business Value of BIM in North America*. Available at: https://bimforum.org/wp-content/uploads/2012/12/MHC-Business-Value-of-BIM-in-North-America-2007-2012-SMR.pdf
McGraw-Hill Construction. (2014). *The Business Value of BIM for Construction in Major Global Markets: How Contractors around the World Are Driving Innovation with Building Information Modelling*. Available at: https://www.icn-solutions.nl/pdf/bim_construction.pdf
McKinsey. (2016). *Imagining Construction's Digital Future*. Available at: https://www.mckinsey.it/idee/imagining-constructions-digital-future
McKinsey Global Institute (MGI). (2011). *Big Data: The Next Frontier for Innovation, Competition, and Productivity*. Available at: https://www.mckinsey.com/~/media/McKinsey/Business%20Functions/McKinsey%20Digital/Our%20Insights/Big%20data%20The%20next%20frontier%20for%20innovation/MGI_big_data_exec_summary.ashx
McPartland R. (2017). *What is a BIM Execution Plan (BEP)?* Available at: https://www.thenbs.com/knowledge/what-is-a-bim-execution-plan-bep.
McPhee, A. (2013). *What is This Thing Called LOD?* Available at: http://practicalbim.blogspot.hk/2013/03/what-is-this-thing-called-lod.html
Meadati, P., Irizarry, J., and Akhnoukh, A. K. (2010). BIM and RFID integration: A pilot study. *Advancing and Integrating Construction Education, Research and Practice Second International Conference on Construction in Developing Countries (ICCIDC–II)*. August 3–5, 2010, Cairo, Egypt (pp. 570–578).
Mell, P., and Grance, T. (2011). *The NIST Definition of Cloud Computing*. Available at: http://faculty.winthrop.edu/domanm/csci411/Handouts/NIST.pdf
Merschbrock, C., and Rolfsen, C. N. (2016). BIM technology acceptance among reinforcement workers-the case of Oslo airport's terminal 2. *Journal of Information Technology in Construction*, *21*, 1–12.
Misuraca, G., Mureddu, F., and Osimo, D. (2014). Policy-making 2.0: Unleashing the power of big data for public governance. In *Open Government* (pp. 171–188). Springer, New York.
Mitchell, D. (2012). 5D BIM: Creating cost certainty and better buildings. Paper presented in *2012 RICS Cobra Conference*, Las Vegas, NV, September 11–13, 2012.
Monteiro, A., and Martins, J. P. (2013). A survey on modelling guidelines for quantity takeoff-oriented BIM-based design. *Automation in Construction*, *35*, 238–253.
Morris, P. W. G., and Geraldi, J. (2011). Managing the institutional context for projects. *Project Management Journal*, *42*(6), 20–32.
Murdoch, T. B., and Detsky, A. S. (2013). The inevitable application of big data to health care. *JAMA*, *309*(13), 1351–1352.
Nagalingam, G., Jayasena, H. S., and Ranadewa, K. A. T. O. (2013). Building information modelling and future quantity surveyor's practice in Sri Lankan construction industry. In *Second World Construction Symposium*, Sri Lanka, June 14–15, 2013. Mawatha: University of Mawatha (pp. 81–92).
Nassar, K. (2011). Assessing building information modelling estimating techniques using data from the classroom. *Journal of Professional Issues in Engineering Education and Practice*, *138*(3), 171–180.

NATSPEC. (2011). *National BIM Guide*. Available at: https://bim.natspec.org/documents/natspec-national-bim-guide

National Building Specification (NBS). (2016). *International BIM Report 2016*. Available at: https://www.thenbs.com/knowledge/nbs-international-bim-report-2016.

National Institute of Building Sciences (NIBS). (2015). *National BIM Standard-United States*. Available at: https://www.nationalbimstandard.org/

Neelamkavil, J., and Ahamed, S. S. (2012). *The Return on Investment from BIM-Driven Projects in Construction*. Ottawa: National Research Council Canada, Institute for Research in Construction.

NewsOn6 (2018). *Storage in Big Data Market Size, Global Overview, Gross Margin Analysis, Development Status, Opportunities, Competitive Landscape and Industry Poised for Rapid Growth by Forecast 2022*. Available at: http://www.newson6.com/story/38295998/storage-in-big-data-market-size-global-overview-gross-margin-analysis-development-status-opportunities-competitive-landscape-and-industry-poised-for

Ohm, P. (2009). Broken promises of privacy: Responding to the surprising failure of anonymization. *UCLA Law Review, 57*, 1701.

Olatunji, O. (2011). Modelling organisations' structural adjustment to BIM adoption: A pilot study on estimating organisations. *Journal of Information Technology in Construction, 16*, 652–668.

Olatunji, O. A., Sher, W., and Gu, N. (2010). Building information modelling and quantity surveying practice. *Emirates Journal for Engineering Research, 15*(1), 67–70.

Padhy, R. P. (2013). Big data processing with Hadoop-MapReduce in cloud systems. *International Journal of Cloud Computing and Services Science, 2*(1), 16.

Pearce, D. (2003). *The Social and Economic Value of Construction: The Construction Industry's Guide to Sustainable Development 2003*. Available at: www.crisp-uk.org.uk/reports/SocialandEconomicValue_FR03.pdf

Pittard, S., and Sell, P. (2016). *BIM and Quantity Surveying*. Abingdon: Routledge.

Poleto, T., de Carvalho, V. D. H., and Costa, A. P. C. S. (2015). The roles of big data in the decision-support process: An empirical investigation. In *International Conference on Decision Support System Technology* (pp. 10–21). Cham: Springer.

Porter, M. E. (1980). *Competitive Strategy: Techniques for Analyzing Industries and Competitors*. New York: The Free Press.

Porter, M. E. (1985). *Competitive Advantage: Creating and Sustaining Superior Performance*. New York: The Free Press.

Pratt, M. J. (2004). Extension of ISO 10303, the STEP standard, for the exchange of procedural shape models. In *Proceedings of Shape Modelling Applications*. 317–326. IEEE.

Project Management Institute (PMI). (2013). *A Guide to the Project Management Body of Knowledge (PMBOK® Guide)*. Hoboken, NJ: Project Management Institute.

Qian, A. Y. (2012). *Benefits and ROI of BIM for Multi-disciplinary Project Management*. Undergraduate dissertation, National University of Singapore.

Raj, E. D., Babu, L. D., Ariwa, E., Nirmala, M., and Krishna, P. V. (2015). Forecasting the trends in cloud computing and its impact on future IT business. In *Cloud Technology: Concepts, Methodologies, Tools, and Applications* (pp. 2354–2372). IGI Global.

Rowlinson, S., and Mcdermott, P. (1999). *Procurement Systems: A Guide to Best Practice in Construction*. London: E&FN SPON.

Royal Institution of Chartered Surveyors (RICS). (1970). *Surveyors and Their Future*. London: Report to the RICS.

Royal Institution of Chartered Surveyors (RICS). (1998). *The APC Requirement and Competencies*. London: Report to the RICS.

Royal Institution of Chartered Surveyors (RICS). (2014). *Overview of a 5D BIM Project*. Available at: http://www.rics.org/Documents/Overview_of_5D_BIM_project_1st_edition_PGguidance_2014.pdf

Royal Institution of Chartered Surveyors (RICS). (2016). *CPD Requirements and Obligations*. Available at: https://www.rics.org/us/regulation1/compliance1/continuing-professional-development-cpd/cpd-requirements-and-obligations/

Russom, P. (2011). Big data analytics. *TDWI Best Practices Report*, *19*(4), 1–34.

Sabol, L. (2008). *Challenges in Cost Estimating with Building Information Modelling*. Paper presented in the *International Facility Management Association (IFMA) World Workplace 2008*, Dallas, October 15–17, 2008.

Saxon, R. (2016). *BIM for Construction Clients*. London: RIBA Publishing.

Schatz, D., Bashroush, R., and Wall, J. (2017). Towards a more representative definition of cyber security. *Journal of Digital Forensics, Security and Law*, *12*(2), 8.

Schlueter, A., and Thesseling, F. (2009). Building information model based energy/exergy performance assessment in early design stages. *Automation in Construction*, *18*(2), 153–163.

Sen, P. K., and Singer, M. J. (1993). *Large Sample Method in Statistics*. Chapman and Hall/CRC.

Shen, W., Hao, Q., and Xue, Y. (2012). A loosely coupled system integration approach for decision support in facility management and maintenance. *Automation in Construction*, *25*, 41–48.

Shih, N. J., and Huang, S. T. (2006). 3D scan information management system for construction management. *Journal of Construction Engineering and Management*, *132*(2), 134–142.

Simon, H. A. (1982). *Models of Bounded Rationality: Behavioral Economics and Business Organization, vol. 2*. The Massachusetts Institute of Technology.

Sinclair, D. (2013). *RIBA Plan of Work 2013 Overview*. London: Royal Institute of British Architects.

Singh, V., Gu, N., and Wang, X. (2011). A theoretical framework of a BIM-based multi-disciplinary collaboration platform. *Automation in Construction*, *20*(2), 134–144.

Sive, T. (2009). *Integrated Project Delivery: Reality and Promise*. Alexandria, VA: Society for Marketing Professional Services Foundation.

Skitmore, R. M., and Marston, V. (1999). *Cost Modelling*. Oxford: Taylor & Francis.

Smith, P. (2014). BIM & the 5D project cost manager. *Procedia-Social and Behavioral Sciences*, *119*, 475–484.

Smith, P. (2016). Project cost management with 5D BIM. *Procedia-Social and Behavioral Sciences*, *226*, 193–200.

Söderlund, J. (2011). Theoretical foundations of project management: Suggestions for a pluralistic understanding. In P. Morris, J. Pinto, and J. Söderlund (eds), *The Oxford Handbook of Project Management* (pp. 37–64). Oxford, England: Oxford University Press.

Solihin, W., and Eastman, C. (2015). Classification of rules for automated BIM rule checking development. *Automation in Construction*, *53*, 69–82.

Song, S., Yang, J., and Kim, N. (2012). Development of a BIM-based structural framework optimisation and simulation system for building construction. *Computers in Industry*, *63*(9), 895–912.

Southwell, J. (1970). *Building Cost Forecasting*. Selected papers on a systematic approach to forecasting building cost. London: RICS Publications.

Stanley, R., and Thurnell, D. P. (2014). The benefits of, and barriers to, implementation of 5D BIM for quantity surveying in New Zealand. *Construction Economics and Building*, *14*(1), 105–117.

Staub-French, S., Fischer, M., Kunz, J., and Paulson, B. (2003a). An ontology for relating features with activities to calculate costs. *Journal of Computing in Civil Engineering, 17*(4), 243–254.

Staub-French, S., Fischer, M., Kunz, J., and Paulson, B. (2003b). A generic feature-driven activity-based cost estimation process. *Advanced Engineering Informatics, 17*(1), 23–39.

Staub-French, S., Forgues, D., Iordanova, I., Kassaian, A., Abdulaal, B., Samilski, M., Cavka, H. B., and Nepal, M. (2011). Building Information Modelling (BIM): An Investigation of "Best Practices" Through Case Studies at Regional, National and International Levels. Available at: http://bim-civil.sites.olt.ubc.ca/files/2014/06/BIMBestPractices2011.pdf

Steel, J., Drogemuller, R., and Toth, B. (2012). Model interoperability in building information modelling. *Software & Systems Modelling, 11*(1), 99–109.

Stewart, R. D., Wyskida, R. M., and Johannes, J. D. (Eds.). (1995). *Cost Estimator's Reference Manual* (Vol. 15). Hoboken, NJ: John Wiley & Sons.

Succar, B. (2009). Building information modelling framework: A research and delivery foundation for industry stakeholders. *Automation in Construction, 18*(3), 357–375.

Sutrisna, M., Buckley, K., Potts, K., and Proverbs, D. (2005). *A Decision Support Tool for the Valuation of Variations on Civil Engineering Projects*. RICS Research Paper Series. London: RICS Publication.

Tang, P., Huber, D., Akinci, B., Lipman, R., and Lytle, A. (2010). Automatic reconstruction of as-built building information models from laser-scanned point clouds: A review of related techniques. *Automation in Construction, 19*(7), 829–843.

Tatum, C. B. (1983). Issues in professional construction management. *Journal of Construction Engineering and Management, 109*(1), 112–119.

Taylor, J. E., and Bernstein, P. G. (2009). Paradigm trajectories of building information modelling practice in project networks. *Journal of Management in Engineering, 25*(2), 69–76.

Teicholz, P. (2004). Labor productivity declines in the construction industry: Causes and remedies. *AECbytes Viewpoint, 4*(14).

Teicholz, P. (2013). *BIM for Facility Managers*. Hoboken, NJ: John Wiley & Sons.

Teizer, J., Caldas, C. H., and Haas, C. T. (2007). Real-time three-dimensional occupancy grid modeling for the detection and tracking of construction resources. *Journal of Construction Engineering and Management, 133*(11), 880–888.

Thurairajah, N., and Goucher, D. (2013). Advantages and challenges of using BIM: A cost consultant's perspective. In *Proceedings of 49th ASC Annual International Conference*, San Luis Oispo, CA, April 2013. San Luis Obispo, CA: Associated Schools of Construction (pp. 1–8).

Towey, D. (2012). *Construction Quantity Surveying: A Practical Guide for the Contractor's QS*. Hoboken, NJ: John Wiley & Sons.

Tuomi, I. (1999). Data is more than knowledge: Implications of the reversed knowledge hierarchy for knowledge management and organizational memory. In *Proceedings of the 32nd Annual Hawaii International Conference on Systems Sciences* (pp. 12-pp). IEEE.

Turkan, Y., Bosche, F., Haas, C. T., and Haas, R. (2012). Automated progress tracking using 4D schedule and 3D sensing technologies. *Automation in Construction, 22*, 414–421.

University of Pittsburgh Medical Center (UPMC). (2017). *Pitt Public Health to Present Big Data Resources at National Meeting*. Available at: http://www.upmc.com/media/NewsReleases/2017/Pages/van-nostrand-informatics.aspx

Vance, D. (1997). Information, knowledge and wisdom: The epistemic hierarchy and computer-based information systems. In *Proceedings of AMCIS 1997*. 165.

Vass, S., and Gustavsson, T. K. (2017). Challenges when implementing BIM for industry change. *Construction Management and Economics, 35*(10), 597–610.

Wang, K. C., Wang, W. C., Wang, H. H., Hsu, P. Y., Wu, W. H., and Kung, C. J. (2016). Applying building information modelling to integrate schedule and cost for establishing construction progress curves. *Automation in Construction*, *72*, 397–410.

Wang, Y., Wang, X., Wang, J., Yung, P., and Jun, G. (2013). Engagement of facilities management in design stage through BIM: Framework and a case study. *Advances in Civil Engineering*, 1–8.

Watson, I. (1999). Case-based reasoning is a methodology not a technology. *Knowledge-Based Systems*, *12*(5–6), 303–308.

Weng, W. H., and Weng, W. T. (2013). Forecast of development trends in big data industry. In *Proceedings of the Institute of Industrial Engineers Asian Conference 2013*, Taipei, July 18–20. Singapore: Springer (pp. 1487–1494).

Weygant, R. S. (2011). *BIM Content Development: Standards, Strategies, and Best Practices*. Hoboken: John Wiley & Sons.

Winch, G. (1989). The construction firm and the construction project: A transaction cost approach. *Construction Management and Economics*, *7*(4), 331–345.

Wolstenholme, A., Austin, S. A., Bairstow, M., Blumenthal, A., Lorimer, J., McGuckin, S., and Guthrie, W. (2009). *Never Waste A Good Crisis: A Review of Progress Since Rethinking Construction and Thoughts for Our Future*. Constructing Excellence.

Wong, A. K., Wong, F. K., and Nadeem, A. (2010). Attributes of building information modelling implementations in various countries. *Architectural Engineering and Design Management*, 6(4), 288–302.

Wong, J. K. W., and Zhou, J. (2015). Enhancing environmental sustainability over building life cycles through green BIM: A review. *Automation in Construction*, *57*, 156–165.

Wong, J., Wang, X., Li, H., and Chan, G. (2014a). A review of cloud-based BIM technology in the construction sector. *Journal of Information Technology in Construction*, *19*, 281–291.

Wong, P. F., Salleh, H., and Rahim, F. A. (2014b). The relationship of building information modelling (BIM) capability in quantity surveying practice and project performance. *International Journal of Civil, Environmental, Structural, Construction and Architectural Engineering*, 8(10), 1039–1044.

World Economic Forum. (2012). *Big Data, Big Impact: New Possibilities for International Development*. Available at: www3.weforum.org/docs/WEF_TC_ MFS_BigDataBig Impact_Briefing_2012.pdf

Woo, J., Wilsmann, J., and Kang, D. (2010). Use of as-built building information modelling. In *Proceedings of Construction Research Congress 2010*, Alberta, May 8–10, 2010. Teston, VA: American Society of Civil Engineers (pp. 538–547).

Woodward, D. G. (1997). Life cycle costing—Theory, information acquisition and application. *International Journal of Project Management*, *15*(6), 335–344.

Woudhuysen, J., and Abley, I. (2004). *Why is Construction So Backward?* London: John Wiley & Sons.

Wu, S., Wood, G., Ginige, K., and Jong, S. W. (2014). A technical review of BIM based cost estimating in UK quantity surveying practice, standards and tools. *Journal of Information Technology in Construction*, *19*, 534–562.

Xie, H., Shi, W., and Issa, R. R. (2011). Using RFID and real-time virtual reality simulation for optimisation in steel construction. *Journal of Information Technology in Construction*, *16*, 291–308.

Xue, F., Lu, W., and Chen, K. (2018a). Automatic generation of semantically rich as-built building information models using 2D images: A derivative-free optimization approach. *Computer-Aided Civil and Infrastructure Engineering*, 1–17. Doi: 10.1111/mice.12378

Xue, F., Chen, K., Lu, W., Niu, Y., and Huang, G. Q. (2018b). Linking radio-frequency identification to Building Information Modeling: Status quo, development trajectory and guidelines for practitioners. *Automation in Construction*, 93, 241–251.

Yan, H., and Demian, P. (2008). Benefits and barriers of building information modelling. In *Proceedings of the 12th International Conference on Computing in Civil and Building Engineering & 2008 International Conference on Information Technology in Construction.* Beijing: Tingshua University Press.

Zaslavsky, A., Perera, C., and Georgakopoulos, D. (2013). *Sensing as a Service and Big Data.* Ithaca, NY: arXiv preprint.

Zeiss, G. (2013). *Widespread Adoption of BIM by National Governments.* Available at: http://geospatial.blogs.com/geospatial/2013/07/widespread-adoption-of-bim-by-national-governments.html

Zheng, L., Lu, W., Chen, K., Chau, W. K., and Niu, Y. (2017) Benefit sharing for BIM implementation: Tackling the moral hazard dilemma in inter-firm cooperation. *International Journal of Project Management*, 35(3), 393–405.

Zhou, L., Perera, S., Udeaja, C., and Paul, C. (2012). Readiness of BIM: A case study of a quantity surveying organisation. In *First UK Academic Conference on BIM*, Northumbria University, UK.

Zhu, J., Zhuang, E., Fu, J., Baranowski, J., Ford, A., and Shen, J. (2016). A framework-based approach to utility big data analytics. *IEEE Transactions on Power Systems*, *31*(3), 2455–2462.

Zikopoulos, P., and Eaton, C. (2011). *Understanding Big Data: Analytics for Enterprise Class Hadoop and Streaming Data.* Bangor, ME: McGraw-Hill Osborne Media.

Index